NATURALIST

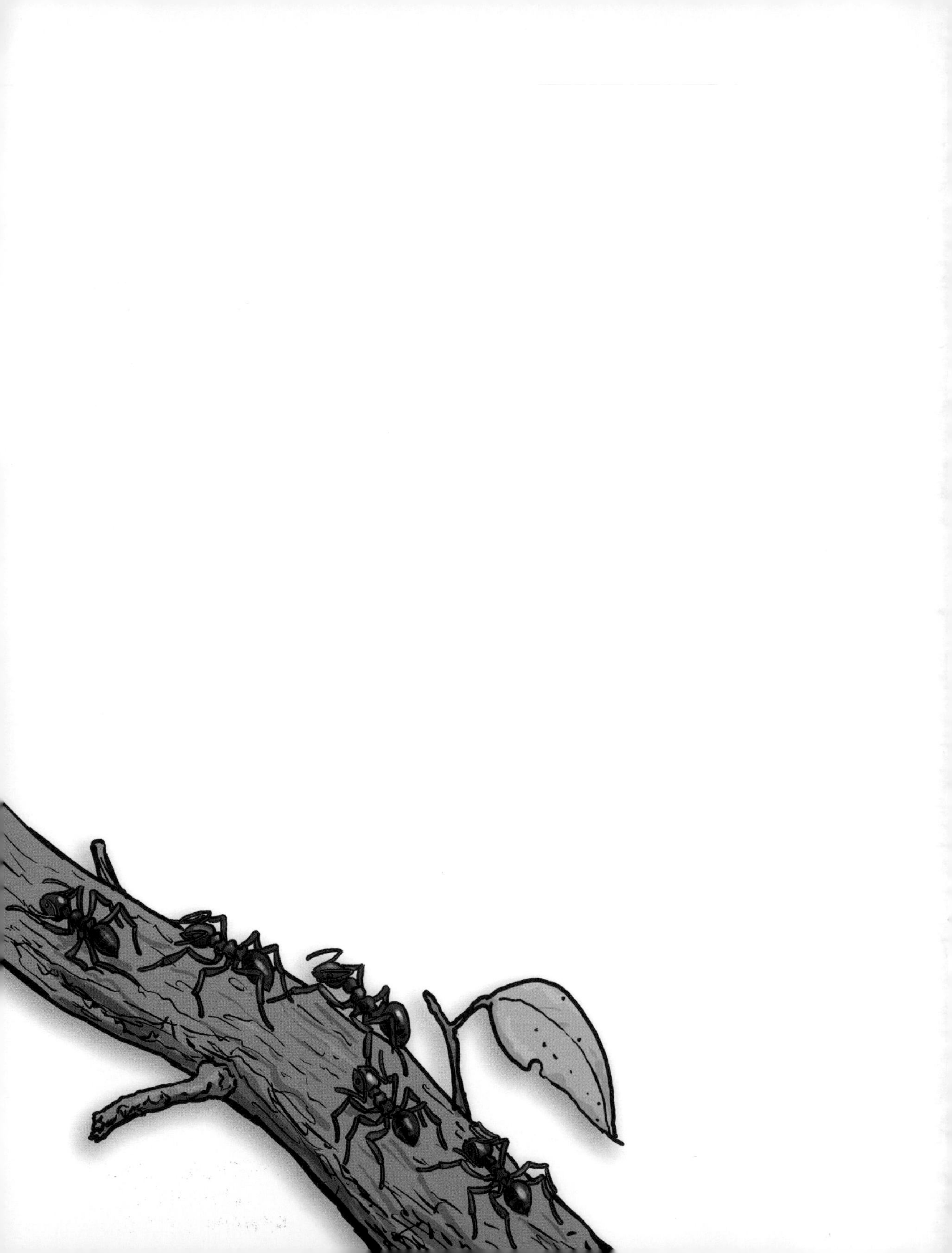

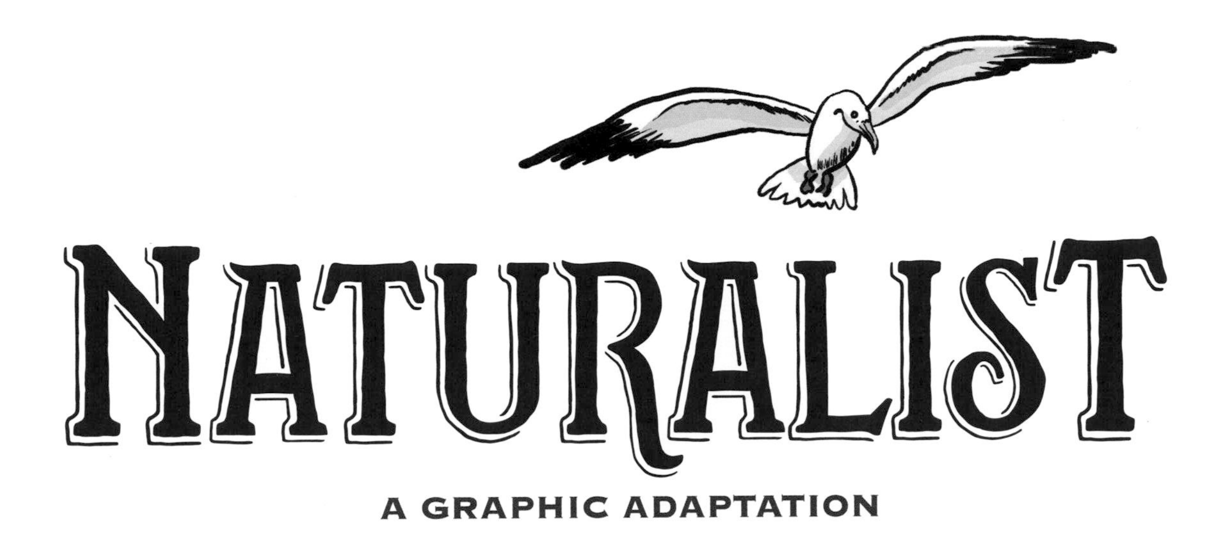

NATURALIST

A GRAPHIC ADAPTATION

EDWARD O. WILSON

ADAPTED BY JIM OTTAVIANI

ART BY C.M.BUTZER

COLORING BY HILARY SYCAMORE

ISLANDPRESS | Washington | Covelo

Library of Congress Control Number: 2020941017

All Island Press books are printed on environmentally responsible materials.

Manufactured in the United States of America
10 9 8 7 6 5 4 3 2 1

Keywords: Ants, Alabama, Biodiversity, biophilia, chemical signaling, conservation, ecology, entomology, evolutionary biology, Florida, Harvard University, island biogeography, National Museum of Natural History, National Zoo, natural history, nature, Pensacola, Pulitzer Prize, Rock Creek Park, speciation, sociobiology, University of Alabama

To the memory of
William L. Brown and Frank M. Carpenter,
great Harvard entomologists
whose support
made my career possible.

FOR ME, NATURE, WITH A CAPITAL N, HOLDS TWO MEANINGS.
IN THE PREVIOUS CENTURY, WHEN I WAS BORN, PEOPLE COULD STILL EASILY THINK OF THEMSELVES AS TRANSCENDENT...
DARK ANGELS CONFINED TO EARTH, AWAITING REDEMPTION BY EITHER SOUL OR INTELLECT.
NOW, MOST OF THE RELEVANT EVIDENCE FROM SCIENCE POINTS IN THE OPPOSITE DIRECTION.
WE WERE BORN IN THE NATURAL WORLD...
WE EVOLVED THERE, STEP BY STEP ACROSS MILLIONS OF YEARS...

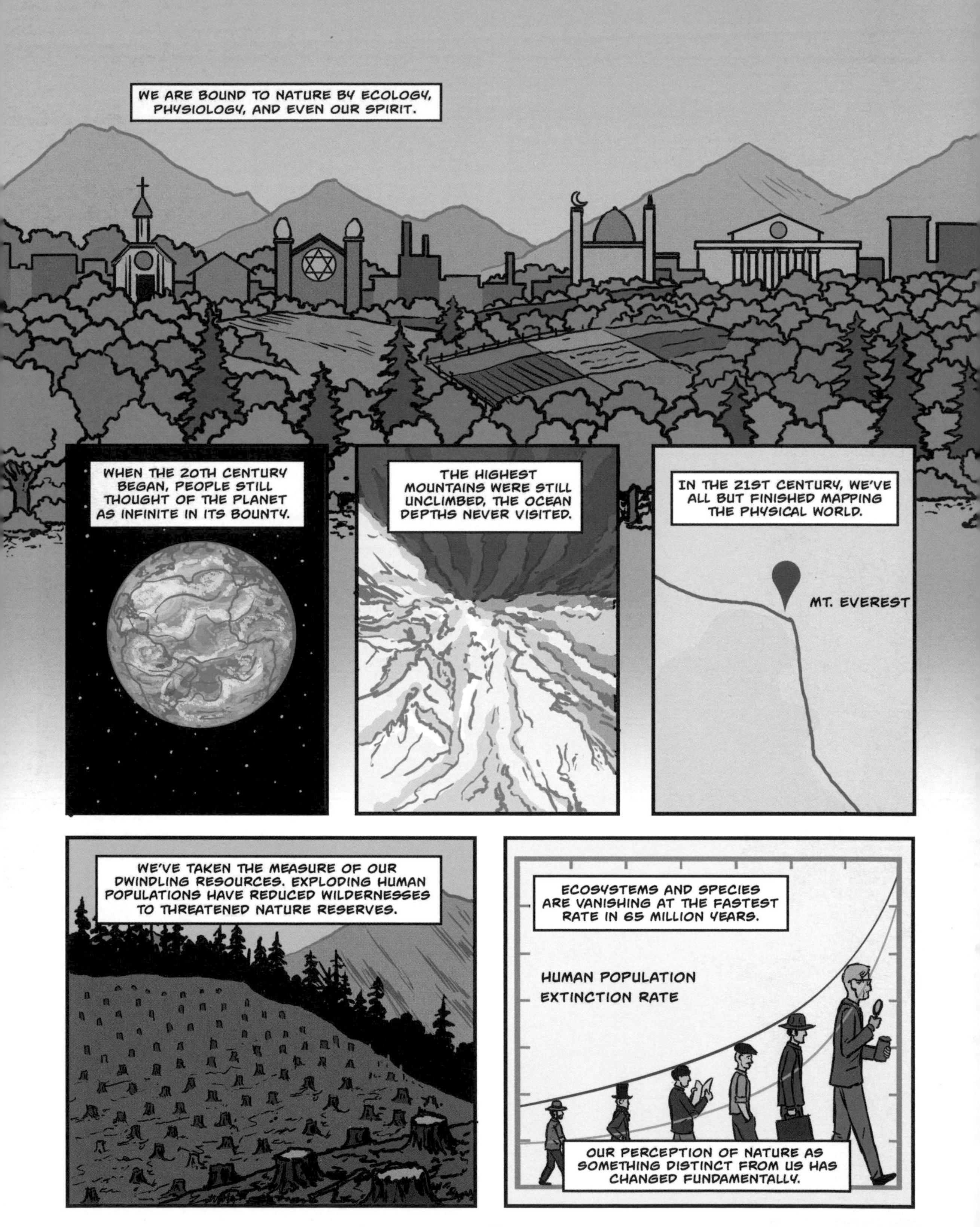
WE ARE BOUND TO NATURE BY ECOLOGY, PHYSIOLOGY, AND EVEN OUR SPIRIT.
WHEN THE 20TH CENTURY BEGAN, PEOPLE STILL THOUGHT OF THE PLANET AS INFINITE IN ITS BOUNTY.
THE HIGHEST MOUNTAINS WERE STILL UNCLIMBED, THE OCEAN DEPTHS NEVER VISITED.
IN THE 21ST CENTURY, WE'VE ALL BUT FINISHED MAPPING THE PHYSICAL WORLD.
MT. EVEREST
WE'VE TAKEN THE MEASURE OF OUR DWINDLING RESOURCES. EXPLODING HUMAN POPULATIONS HAVE REDUCED WILDERNESSES TO THREATENED NATURE RESERVES.
ECOSYSTEMS AND SPECIES ARE VANISHING AT THE FASTEST RATE IN 65 MILLION YEARS.
HUMAN POPULATION
EXTINCTION RATE
OUR PERCEPTION OF NATURE AS SOMETHING DISTINCT FROM US HAS CHANGED FUNDAMENTALLY.

NATURALIST

PART ONE

Daybreak in Alabama

When I get to be a composer
I'm gonna write me some music about
Daybreak in Alabama
And I'm gonna put the purtiest songs in it
Rising out of the ground like swamp mist
And falling out of heaven like soft dew.

[Langston Hughes]

SO, COME BACK WITH ME TO PENSACOLA, OCTOBER 1935.
GRAPSIDAE
Grapsidae

LET'S GO FOR A WALK.
SPLOINK!

Chil
RESTAUR

FWSSSST!
A MIRACLE OF MODERN TECHNOLOGY!
THANKS, SON.
Childs
RESTARAUNT
OKAY, COME ON. THERE'S LOTS TO SEE.
GOOD STUFF, RIGHT?
YEAH. ERROL FLYNN SKEWERS BASIL RATHBONE IN A DUEL ON THE TREASURE ISLAND.
COOLED BY ICED AIR
BE CAREFUL WHEN YOU CROSS ROMANA STREET UP AHEAD. A KID ON A BIKE GOT RUN OVER THERE LAST YEAR.
YEAH SURE. I HEARD ABOUT THAT.

ALONG, AT THE COURTHOUSE, IS SOMETHING MORE INTERESTING THAN A KID GETTING RUN OVER.
FIRE ANTS!
SOLENOPSIS GEMINATA

PENSACOLA HAD ALL KINDS OF GREAT STUFF, INCLUDING BOTH MY PARENTS.
HOME
BIKES AND CARS
FIRE ANTS
AUTOMATIC DOORS
FLASH GORDON, CAPTAIN BLOOD, AIR-CONDITIONING
GRAPSID CRABS
SHRIEK! SHRIEK! SHRIEK!
CYANOCITTA CRISTATA
QUERCUS GEMINATA
LOOK DOWN AND YOU SEE LION ANTS.
SHRIEK!
SHRIEK!
SHRIEK!
LOREM IPSUM
DORYMYRMEX
CRUSH ONE, AND THE UNMISTAKABLE SMELL OF A DOLICHODERINE ANT HITS YOUR NOSE.
SHRIE
SHRIEK!
HEPTANONE
METHYLHEPTENONE
I CAN TELL YOU NOW THAT THE ODOR COMES FROM THE PYGIDIAL GLANDS. WORKERS SECRETE IT TO DEFEND THE COLONY AGAINST ENEMIES AND ALARM NESTMATES TO APPROACHING DANGER.
PYGIDIAL GLAND

I RETURN TO THIS SPOT AS OFTEN AS I CAN.
YOU NEED HELP THERE, SON?

PHEIDOLE DENTATA

NO SIR, JUST... JUST WOOL GATHERING, I SUPPOSE.

I HATE TO BREAK IT TO YOU, BUT THAT AIN'T WOOL.

HAH. TRUE ENOUGH. I USED TO LIVE HERE. AND BACK THEN, I WAS INTERESTED IN ANTS.
I'M STILL INTERESTED IN ANTS.

WELL, GOT SOME INTERESTING ONES DOWN WHERE I LIVE. C'MON BY SOMETIME.

I'LL KEEP COMING HERE WHENEVER I VISIT PENSACOLA, TO SEE IF THE DORYMYRMEX HAVE RETURNED TO THIS SPECIAL SPACE.

SO FAR, MY SURVEILLANCE HAS LASTED SOME EIGHTY YEARS—IF I'M FORTUNATE, I'LL HAVE A FEW MORE.
YOU WILL THINK THIS A STRANGE JOURNEY AND A STRANGER OBSESSION, BUT I DON'T.

IT'S HOW LONG-TERM MEMORY WORKS. THE MIND WITH A SEARCH IMAGE IS LIKE A BARRACUDA.

IT LOCKS ON A SIGNAL, RUSHES FORWARD, AND...

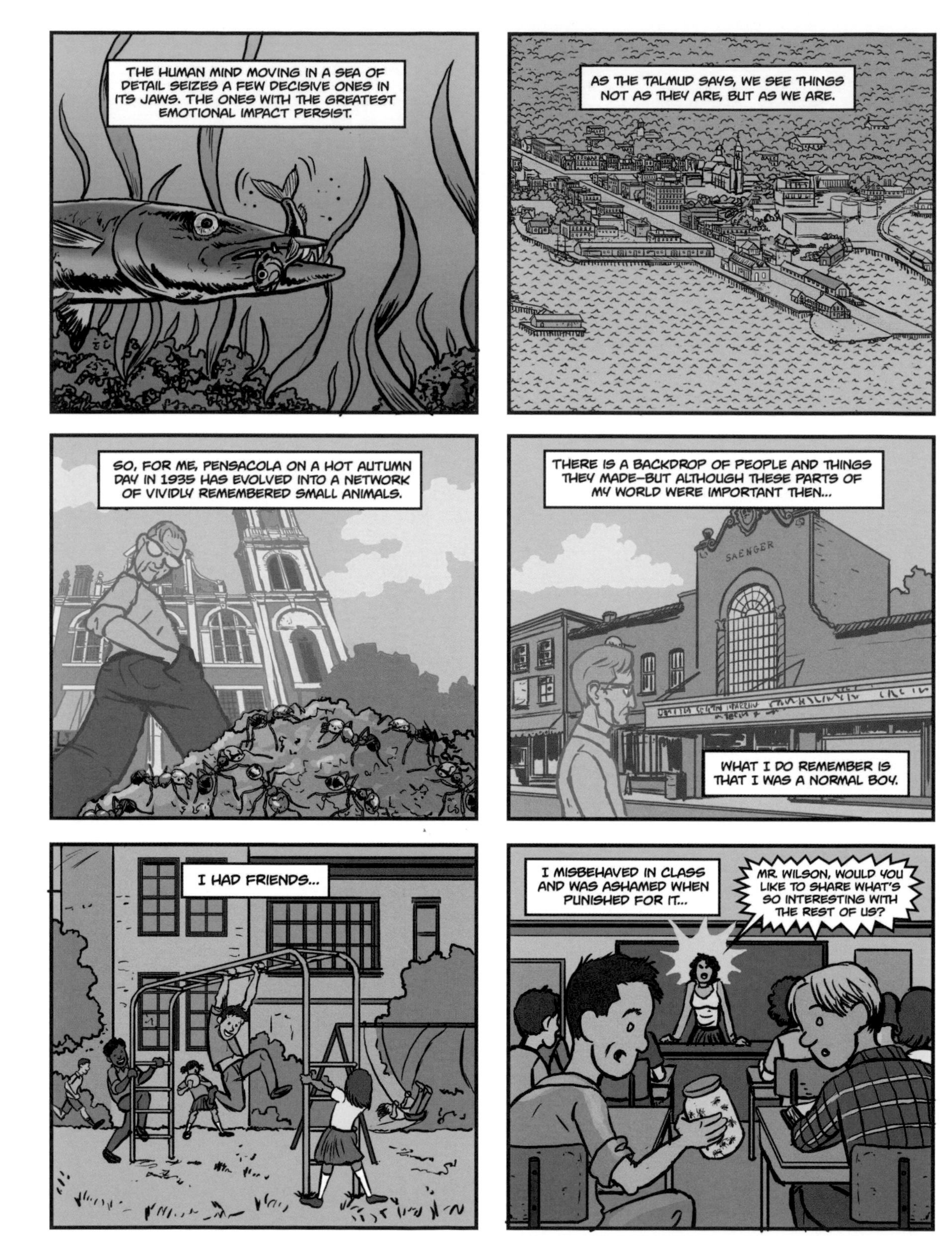
THE HUMAN MIND MOVING IN A SEA OF DETAIL SEIZES A FEW DECISIVE ONES IN ITS JAWS. THE ONES WITH THE GREATEST EMOTIONAL IMPACT PERSIST.
AS THE TALMUD SAYS, WE SEE THINGS NOT AS THEY ARE, BUT AS WE ARE.
SO, FOR ME, PENSACOLA ON A HOT AUTUMN DAY IN 1935 HAS EVOLVED INTO A NETWORK OF VIVIDLY REMEMBERED SMALL ANIMALS.
THERE IS A BACKDROP OF PEOPLE AND THINGS THEY MADE—BUT ALTHOUGH THESE PARTS OF MY WORLD WERE IMPORTANT THEN...
SAENGER
WHAT I DO REMEMBER IS THAT I WAS A NORMAL BOY.
I HAD FRIENDS...
I MISBEHAVED IN CLASS AND WAS ASHAMED WHEN PUNISHED FOR IT...
MR. WILSON, WOULD YOU LIKE TO SHARE WHAT'S SO INTERESTING WITH THE REST OF US?

I OBEYED MY ELDERS...
(MOSTLY)...

AND I ENJOYED FLORIDA'S WINTER HARVEST...
GOT US SOME CHINQUAPINS!

HEY, THERE'S A BUNCH OF PECANS DOWN OVER HERE.
CARYA ILLINOINENSIS. GOOD EATIN'.

THE YEARS HAVE DRAINED THESE MEMORIES OF MOST OF THEIR IMPORTANCE. FINE DETAILS AND EMOTIONAL FORCE HAVE LARGELY ERODED.

FOR EXAMPLE, I HAVE NO MEMORY OF THE FAMILY I STAYED WITH IN THE SUMMER OF 1936.
I DON'T REMEMBER WHAT THEY LOOKED LIKE, THEIR AGES, OR EVEN HOW MANY THERE WERE. NO NEED TO. BUT THE PLACE?

PARADISE BEACH. IT IS ON THE EAST SHORE OF FLORIDA'S PERDIDO BAY, NOT FAR FROM PENSACOLA.
YOU CAN SEE ALABAMA ACROSS THE WATER.

EACH MORNING, AFTER BREAKFAST, I LEFT THE SMALL SHOREFRONT HOUSE TO WANDER.

BACK IN TIME FOR LUNCH, OUT AGAIN, BACK FOR DINNER.
BLAH
BLAH

OUT ONCE AGAIN...

...AND, FINALLY, OFF TO BED TO RELIVE MY CONTINUING ADVENTURE BEFORE FALLING ASLEEP.

I WONDERED WHERE NEEDLEFISH WENT AT NIGHT, BUT NEVER FOUND OUT.

GOOD EATIN'

AT SEVEN YEARS OLD, EVERY SPECIES, LARGE AND SMALL, WAS A WONDER TO BE EXAMINED...

THOUGHT ABOUT...

AND, IF POSSIBLE, CAPTURED AND EXAMINED AGAIN.
MAYBE NOT

LET ME SAY SOMETHING ABOUT THE PSYCHOLOGY OF MONSTER HUNTING.
Hic Sunt Monstra

I ESTIMATE THAT WHEN I WAS SEVEN YEARS OLD, I SAW ANIMALS AT ABOUT TWICE THE SIZE I SEE THEM NOW.
SO GIANTS EXIST, EVEN IF ADULTS DON'T CHOOSE TO CLASSIFY THEM AS SUCH.
I TELL YOU THIS LITTLE BOY'S STORY OF MEDUSAS, RAYS, AND SEA MONSTERS BECAUSE IT ILLUSTRATES, I THINK, HOW A NATURALIST IS CREATED.
A CHILD COMES TO THE EDGE OF DEEP WATER WITH A MIND PREPARED FOR WONDER.
HANDS-ON EXPERIENCE AT THE CRITICAL TIME, NOT SYSTEMATIC KNOWLEDGE, IS WHAT COUNTS IN THE MAKING OF A NATURALIST.
BETTER TO BE AN UNTUTORED SAVAGE FOR A WHILE.
BETTER THAT THE SUMMER AT PARADISE BEACH WAS NOT AN EDUCATIONAL EXERCISE PLANNED BY ADULTS. BETTER THAT IT WAS AN ACCIDENT IN A HAPHAZARD LIFE.

DURING THAT BRIEF TIME, HOWEVER, A SECOND ACCIDENT OCCURRED.
IT DETERMINED WHAT KIND OF NATURALIST I WOULD EVENTUALLY BECOME.

THE PINFISH CARRIES TEN NEEDLE-LIKE SPINES THAT STICK STRAIGHT UP WHEN IT IS THREATENED.
YANK!
LAGODON RHOMBOIDES

I CARELESSLY YANKED TOO HARD, AND...

ONE OF ITS SPINES PIERCED THE PUPIL OF MY RIGHT EYE.

THE PAIN WAS EXCRUCIATING, AND I SUFFERED FOR HOURS. BUT BEING ANXIOUS TO STAY OUTDOORS...
EDWARD, ARE YOU OKAY?

I CAN'T REMEMBER IF THE HOST FAMILY UNDERSTOOD THE PROBLEM, BUT THEY DIDN'T TAKE ME IN FOR MEDICAL TREATMENT.
I'M FINE.

BUT SEVERAL MONTHS LATER, BACK HOME IN PENSACOLA, THE PUPIL BEGAN TO CLOUD OVER.
IT'S A TRAUMATIC CATARACT. HE NEEDS TO GO TO THE HOSPITAL AND HAVE THE LENS REMOVED.

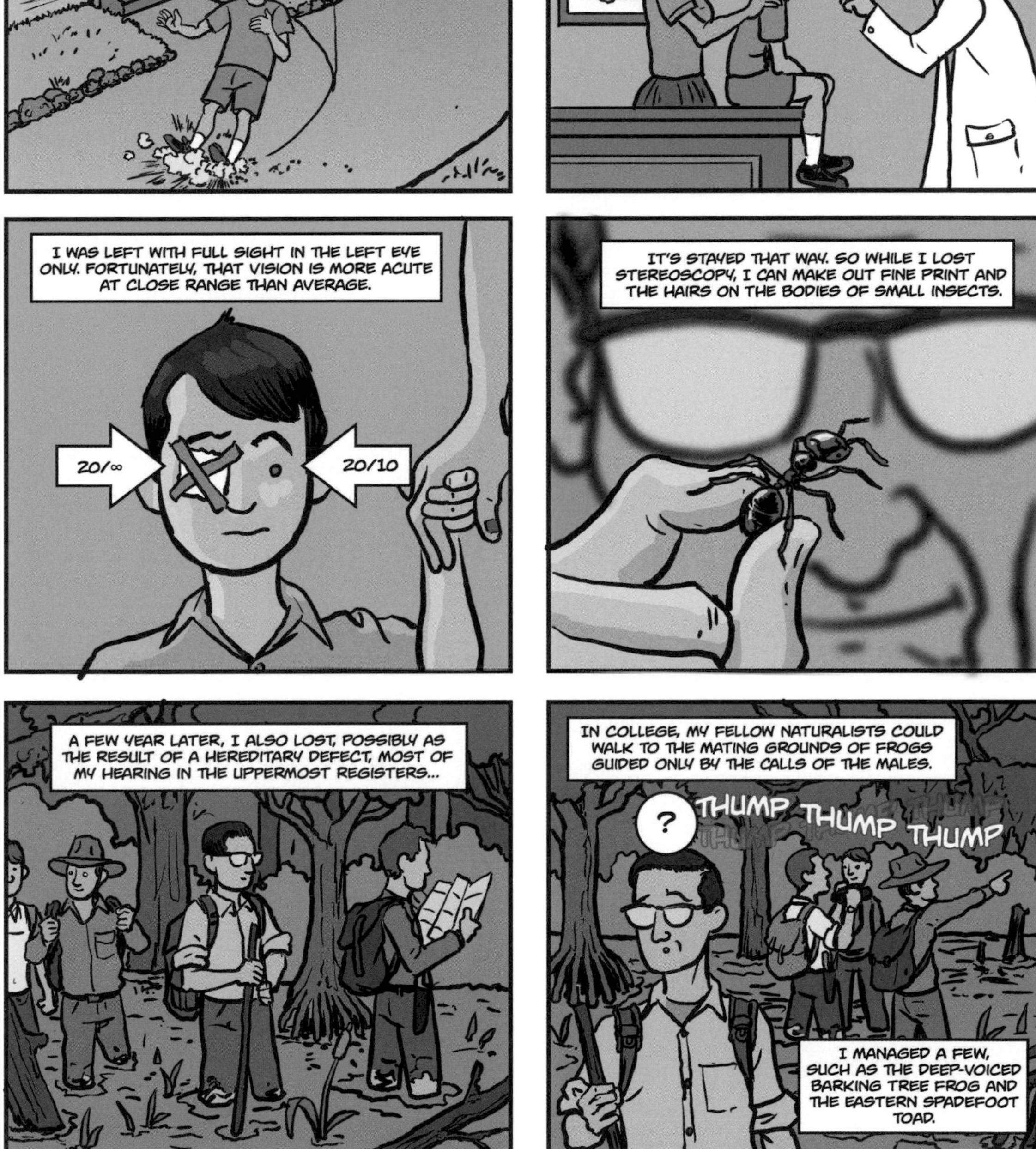
THE NEXT DAY, THE PAIN HAD SUBSIDED INTO MILD DISCOMFORT, AND THEN IT DISAPPEARED.
I WAS LEFT WITH FULL SIGHT IN THE LEFT EYE ONLY. FORTUNATELY, THAT VISION IS MORE ACUTE AT CLOSE RANGE THAN AVERAGE.
20/∞
20/10
IT'S STAYED THAT WAY. SO WHILE I LOST STEREOSCOPY, I CAN MAKE OUT FINE PRINT AND THE HAIRS ON THE BODIES OF SMALL INSECTS.
A FEW YEAR LATER, I ALSO LOST, POSSIBLY AS THE RESULT OF A HEREDITARY DEFECT, MOST OF MY HEARING IN THE UPPERMOST REGISTERS...
IN COLLEGE, MY FELLOW NATURALISTS COULD WALK TO THE MATING GROUNDS OF FROGS GUIDED ONLY BY THE CALLS OF THE MALES.
?
THUMP THUMP THUMP
I MANAGED A FEW, SUCH AS THE DEEP-VOICED BARKING TREE FROG AND THE EASTERN SPADEFOOT TOAD.

BUT FROM MOST SPECIES, ALL I DETECTED WAS A VAGUE BUZZING IN THE EARS.
THIS ALSO MAKES ME A WRETCHED BIRDWATCHER... I CAN'T LOCATE THEM UNLESS THEY OBLIGINGLY FLUTTER PAST IN CLEAR VIEW.
ON THE POSITIVE SIDE, BLUEJAYS AREN'T AS ANNOYING AS THEY USED TO BE.
I THINK I WAS DESTINED TO BECOME AN ENTOMOLOGIST, COMMITTED TO MINUTE, QUIET, CRAWLING, AND FLYING INSECTS...
NOT BECAUSE OF ANY GENIUS. JUST A FORTUITOUS CONSTRICTION OF PHYSIOLOGICAL ABILITY.
I HAD TO HAVE ONE KIND OF ANIMAL IF NOT ANOTHER, BECAUSE THE FIRE HAD BEEN LIT.
I TOOK WHAT I COULD GET, AND TURNED MY SURVIVING EYE TO THE GROUND.
I CELEBRATE THE LITTLE THINGS OF THE WORLD.

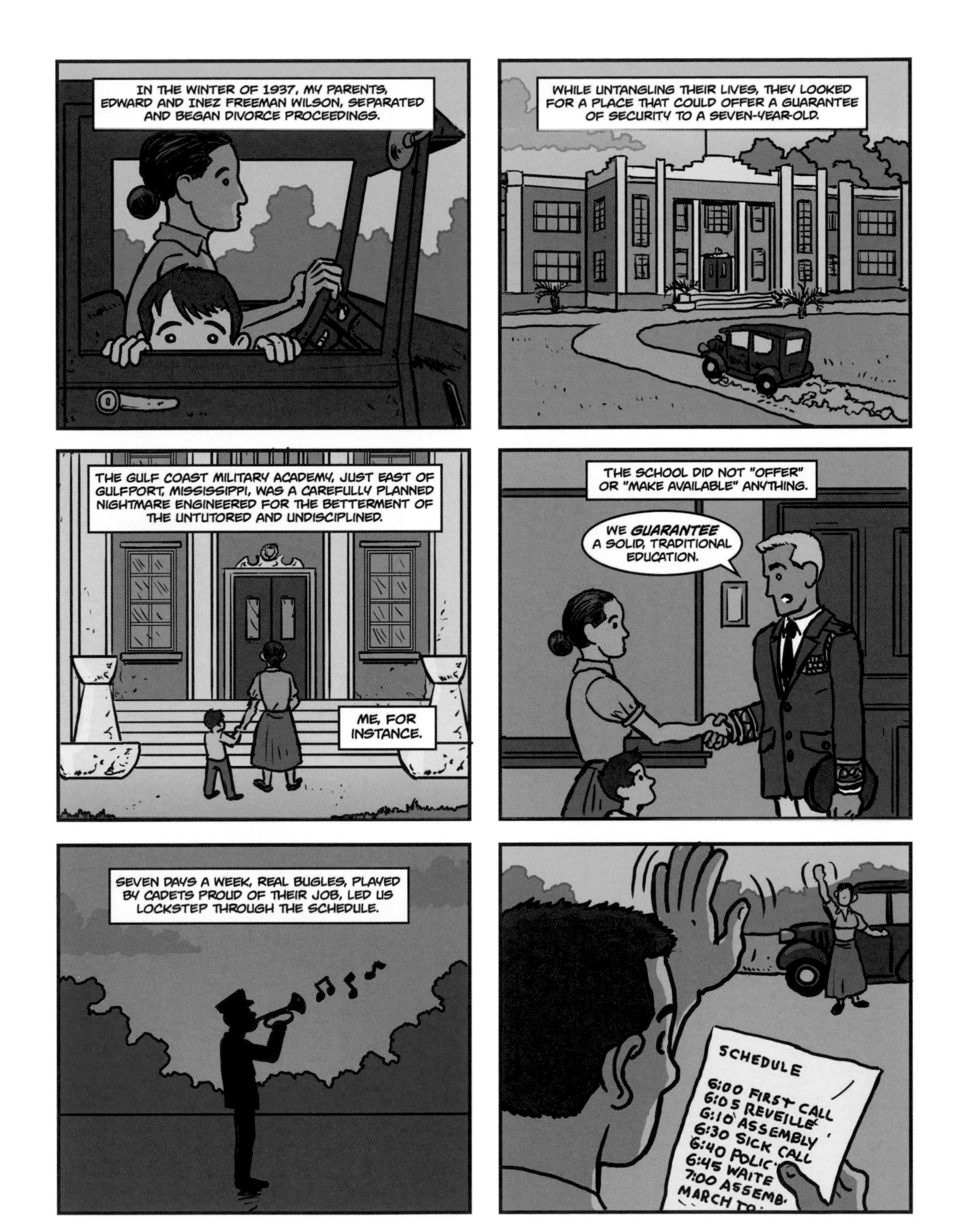
IN THE WINTER OF 1937, MY PARENTS, EDWARD AND INEZ FREEMAN WILSON, SEPARATED AND BEGAN DIVORCE PROCEEDINGS.
WHILE UNTANGLING THEIR LIVES, THEY LOOKED FOR A PLACE THAT COULD OFFER A GUARANTEE OF SECURITY TO A SEVEN-YEAR-OLD.
THE GULF COAST MILITARY ACADEMY, JUST EAST OF GULFPORT, MISSISSIPPI, WAS A CAREFULLY PLANNED NIGHTMARE ENGINEERED FOR THE BETTERMENT OF THE UNTUTORED AND UNDISCIPLINED.
ME, FOR INSTANCE.
THE SCHOOL DID NOT "OFFER" OR "MAKE AVAILABLE" ANYTHING.
WE GUARANTEE A SOLID, TRADITIONAL EDUCATION.
SEVEN DAYS A WEEK, REAL BUGLES, PLAYED BY CADETS PROUD OF THEIR JOB, LED US LOCKSTEP THROUGH THE SCHEDULE.
SCHEDULE
6:00 FIRST CALL
6:05 REVEILLE
6:10 ASSEMBLY
6:30 SICK CALL
6:40 POLIC
6:45 WAITE
7:00 ASSEMB.
MARCH TO

THERE WILL BE ARITHMETIC, ALGEBRA, GEOMETRY, PHYSICS, CHEMISTRY.
SCHOOL CALL WAS AT 7:40. THE CURRICULUM WAS LAID BEFORE US IN RESONANT SINGLE WORDS...

NOT "INTRODUCTION TO CHEMISTRY." CHEMISTRY, FULL STOP.

THERE WILL ALSO BE HISTORY, ENGLISH, AND FOREIGN LANGUAGE.

NO ART, FIELD TRIPS, AND CERTAINLY NOTHING WITH WIMPY TITLES LIKE "THE AMERICAN EXPERIENCE."
AND SO WE TRAMPED FORWARD THROUGH THE DAY, FINALLY TO DINNER AND MORE BUGLES.

6:50 CALL TO QUAR
7:00 STUDY (NO RA
9:15 TATTOO
9:30 TAPS

NO TALKING AFTER TAPS, OR YOU GO ON DELINQUENCY REPORT.
SUCH MISBEHAVIOR BROUGHT TIME IN THE BULLRING...

...AN ACTIVITY NOT MENTIONED IN THE BROCHURES.
WE MARCHED FOR ONE TO SEVERAL HOURS, DEPENDING ON THE SERIOUSNESS OF THE CHARGE.
IT WAS A GOOD TIME TO GET AWAY FROM THE OTHERS AND DAYDREAM.

AS I RECALL, MOST OF MY SINS INVOLVED TALKING WITH OTHER CADETS DURING CLASS.
THE LESSON DID NOT TAKE.
FORMICA LIGNIPERDA!
AS A UNIVERSITY PROFESSOR, I SPENT ALMOST ALL OF MY TIME TALKING IN CLASS.
WEDNESDAY AFTERNOONS WERE FOR FUN, IN THE GCMA WAY OF THINKING...
OKAY WILSON, THAT'LL DO.
...AT LEAST FOR CADETS FREE OF PUNISHMENT. AND I DIDN'T SPEND ALL OF MY TIME AT GCMA TRAVELING IN CIRCLES...
A LOT, BUT NOT ALL.
SO I, TOO, WENT TO GULFPORT FOR MILKSHAKES, MOVIES, AND JUST WALKING AROUND. THIS DOLLOP OF FRIVOLITY WAS ALL WELL AND GOOD, BUT I PINED FOR MY BELOVED GULF OF MEXICO.
IT WAS ALWAYS IN FULL SIGHT FROM THE ACADEMY'S FRONT LAWN, BUT CADETS WERE SENSIBLY FORBIDDEN TO CROSS THE TWO-LANE HIGHWAY THAT SEPARATED THE SCHOOL GROUNDS FROM THE SEAWALL AND BEACH.
GULF COAST MILITARY ACADEMY

IT WAS A LONG SCHOOL YEAR, BUT WHEN I WAS SPRUNG FROM THE GULF COAST MILITARY ACADEMY IN THE SUMMER OF 1937, I CARRIED AN INOCULUM OF THE MILITARY CULTURE.
YOU CAN CALL ME MRS. RAUB, EDWARD.
YES MA'AM. WHERE DID YOU GET THAT? WHAT IS IT?
EDWARD!
SHE WAS THE PERFECT GRANDMOTHER–ATTENTIVE TO MY EVERY NEED AND LISTENING CAREFULLY TO EVERY STORY OF MY LIFE...
AND THAT IS OF COURSE WHEN I KNEW MY...MY *DESTINY.*
YES, I SEE. AND MY, WHAT A VOCABULARY YOU LEARNED AT THAT SCHOOL.
A LIFE I CONSIDERED TO HAVE BEEN BOTH LONG AND FILLED WITH MEANING.

UP TO COLLEGE-AGE ADULT MALES WERE "SIR" AND LADIES "MA'AM," REGARDLESS OF THEIR STATION.
YOU'LL BE STAYING WITH THE RAUBS...FOR NOW. SAY HELLO, EDWARD.
YES MA'AM. HELLO SIR, MA'AM.
IT'S ALRIGHT, INEZ. WHY, E.J. GAVE IT TO ME. DIDN'T YOU, E.J.? IT'S A COUGAR'S CLAW.
WHERE ARE COUGARS FOUND? WHAT DO THEY DO?
SHE TOLD ME. FASCINATING! MONSTERS OF THE LAND.

NOW, LET'S GET OURSELVES CLEANED UP. IT'S ALMOST TIME TO MEET YOUR MOTHER.
SO EVEN WHILE I CONTINUED TO GIVE OUT SIRS AND MA'AMS WITH PLEASURE, I SOON CALLED HER "MOTHER."
AND I GAVE HER NO PROBLEMS WITH MANNERS OR DISCIPLINE.
THE GCMA HAD TAKEN CARE OF THAT. I INSTINCTIVELY RESPECT AUTHORITY AND BELIEVE, EMOTIONALLY IF NOT INTELLECTUALLY, THAT IT SHOULD BE PERTURBED ONLY FOR CONSPICUOUS CAUSE.
AT MY CORE, I AM A SOCIAL CONSERVATIVE, A LOYALIST. I CHERISH TRADITIONAL INSTITUTIONS, THE MORE VENERABLE AND RITUAL-LADEN THE BETTER.
AND I HAVE SPENT A GOOD DEAL OF TIME DURING MY CAREER AS A SCIENTIST THINKING ABOUT THE ORIGINS OF SELF-SACRIFICE AND HEROISM.
I CANNOT SAY I UNDERSTAND THEM FULLY IN HUMAN TERMS.
Onward Christian soldiers, marching as to war
With the cross of Jesus, going on before
BUT THE PEOPLE I FIND IT EASIEST TO ADMIRE ARE THOSE WHO CONCENTRATE ALL THE COURAGE AND SELF-DISCIPLINE THEY POSSESS TOWARD A SINGLE, WORTHY GOAL.

BLEND WITH OURS YOUR VOICES IN THE TRIUMPH SONG.
EXPLORERS, MOUNTAIN CLIMBERS, ULTRAMARATHONERS, MILITARY HEROES, AND A VERY FEW SCIENTISTS...
ONWARD, THEN, YE PEOPLE, JOIN OUR HAPPY THRONG.
SCIENTISTS WHO DRIVE TOWARD DAUNTING GOALS WITH NERVES STEELED AGAINST FAILURE AND A READINESS TO ACCEPT PAIN, AS MUCH TO TEST THEIR OWN CHARACTER AS TO PARTICIPATE IN THE SCIENTIFIC CULTURE.
SCIENCE IS MODERN CIVILIZATION'S HIGHEST ACHIEVEMENT, BUT IT HAS FEW HEROES.

MOTHER RAUB AND E.J. WERE METHODISTS, MEANING THE SERVICES THEY ATTENDED, AFTER DROPPING ME OFF AT THE FIRST BAPTIST CHURCH, WERE MORE SEDATE AND LESS EVANGELICAL.
BUT SHE WAS A FIERCE MORALIST IN THE STRICTER METHODIST TRADITION.
SWEAR TO ME THAT YOU WILL NEVER SMOKE, DRINK, OR GAMBLE.
SURE, I SWEAR.
WHY WOULD ANYONE WANT TO DO THAT STUFF ANYWAY?
IT'S BEEN EASY TO KEEP MY PROMISE–EXCEPT FOR THE ODD GLASS OF WINE OR BEER WITH MEALS–MAINLY BECAUSE I DON'T MUCH CARE FOR THE TASTE OF THE STUFF.
AND, AT A DEEPER LEVEL, BECAUSE OF THE LATER DOWNWARD SPIRAL OF MY FATHER'S LIFE AS A RESULT OF ALCOHOLISM, WHICH I WITNESSED WITH HELPLESS DESPAIR.
BUT, THAT WAS ALL TO COME. THAT YEAR, I ACQUIRED A BLACK CAT AND PLANTED A SMALL GARDEN OF MY OWN IN THE BACKYARD.
HOLE TO CHINA
QUERCUS LAEVIS
CALADIUM
DIOSPYROS VIRGINIANA

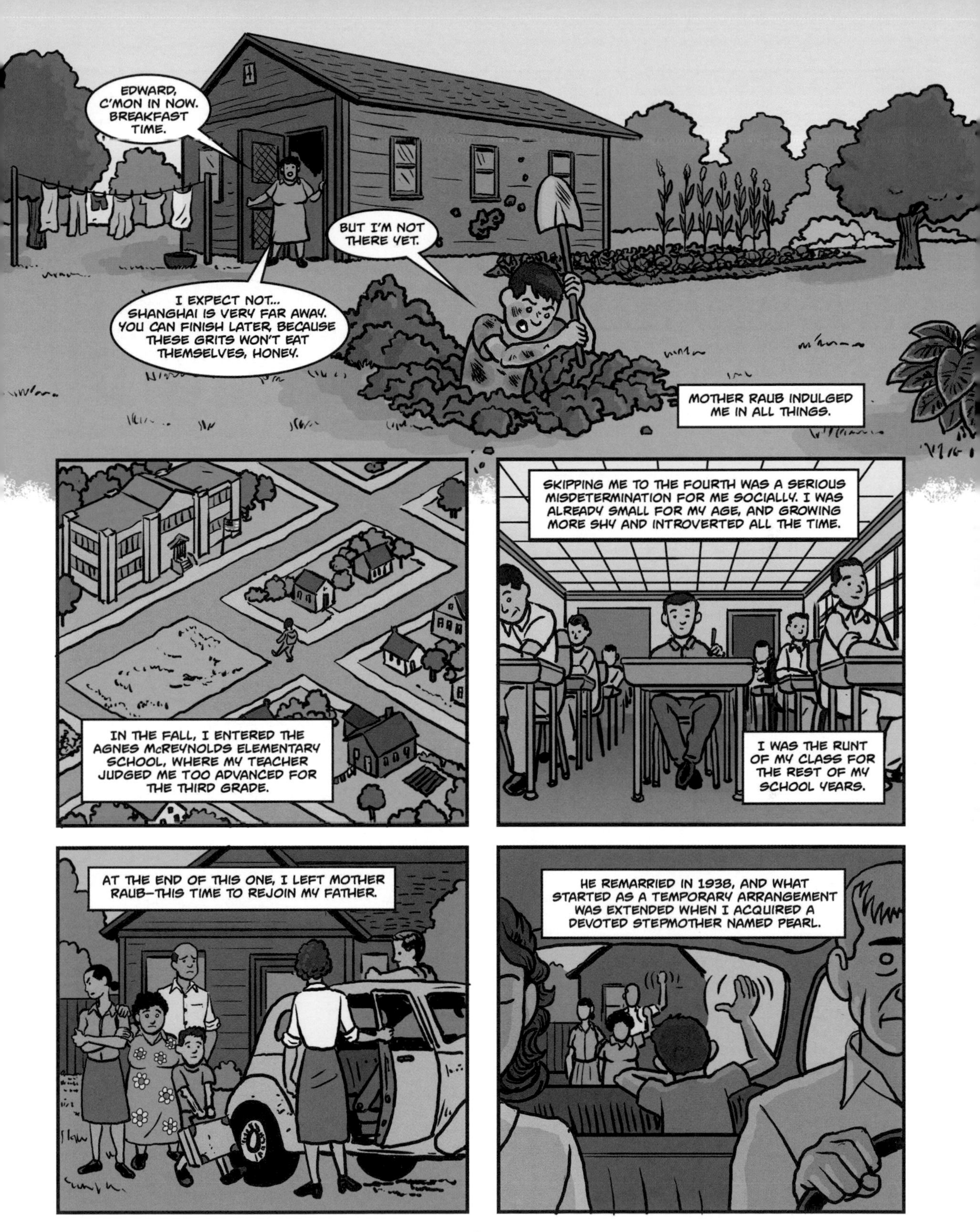
EDWARD, C'MON IN NOW. BREAKFAST TIME.
BUT I'M NOT THERE YET.
I EXPECT NOT... SHANGHAI IS VERY FAR AWAY. YOU CAN FINISH LATER, BECAUSE THESE GRITS WON'T EAT THEMSELVES, HONEY.
MOTHER RAUB INDULGED ME IN ALL THINGS.
IN THE FALL, I ENTERED THE AGNES McREYNOLDS ELEMENTARY SCHOOL, WHERE MY TEACHER JUDGED ME TOO ADVANCED FOR THE THIRD GRADE.
SKIPPING ME TO THE FOURTH WAS A SERIOUS MISDETERMINATION FOR ME SOCIALLY. I WAS ALREADY SMALL FOR MY AGE, AND GROWING MORE SHY AND INTROVERTED ALL THE TIME.
I WAS THE RUNT OF MY CLASS FOR THE REST OF MY SCHOOL YEARS.
AT THE END OF THIS ONE, I LEFT MOTHER RAUB—THIS TIME TO REJOIN MY FATHER.
HE REMARRIED IN 1938, AND WHAT STARTED AS A TEMPORARY ARRANGEMENT WAS EXTENDED WHEN I ACQUIRED A DEVOTED STEPMOTHER NAMED PEARL.

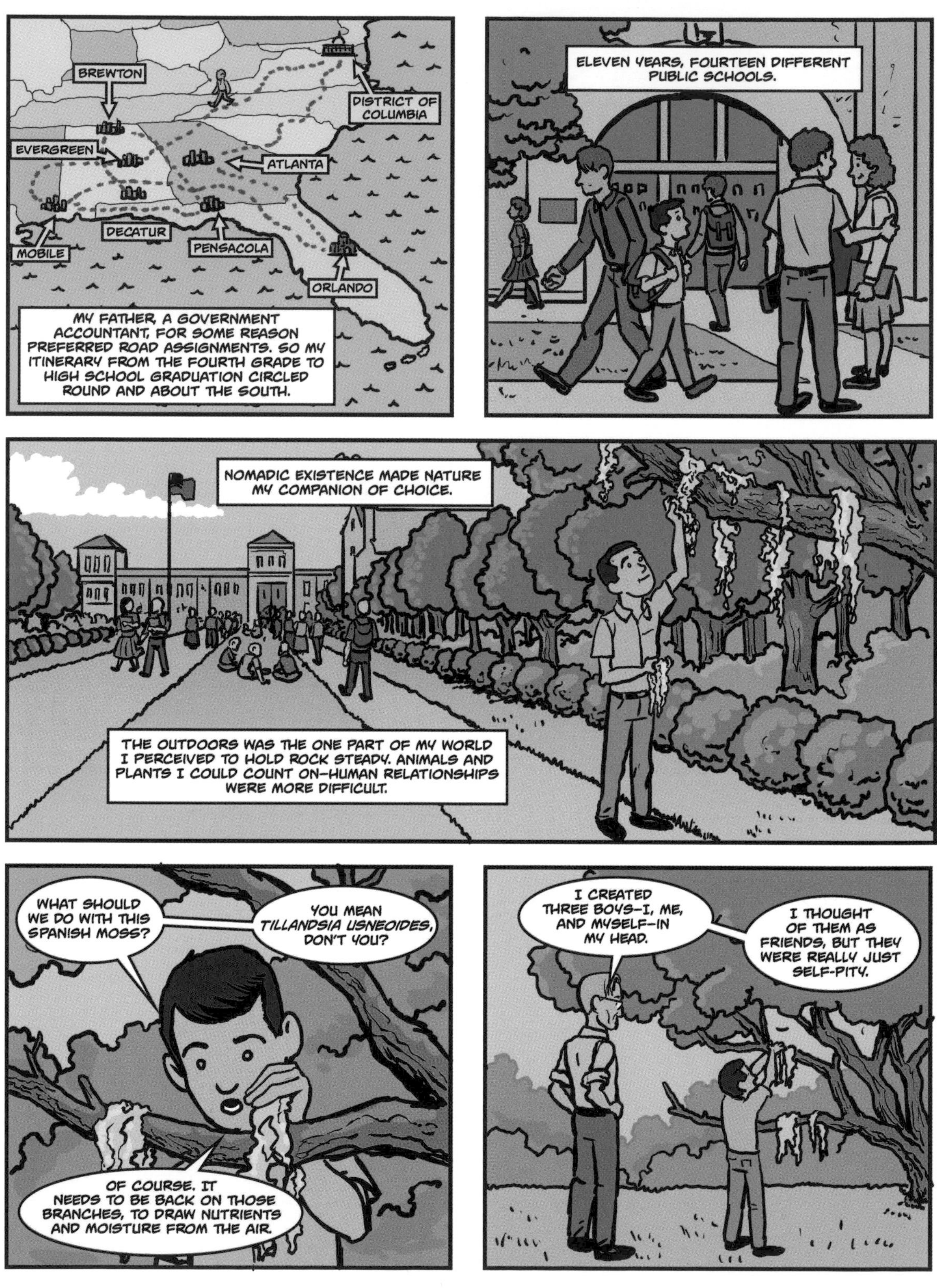
BREWTON
DISTRICT OF COLUMBIA
EVERGREEN
ATLANTA
DECATUR
MOBILE
PENSACOLA
ORLANDO
MY FATHER, A GOVERNMENT ACCOUNTANT, FOR SOME REASON PREFERRED ROAD ASSIGNMENTS. SO MY ITINERARY FROM THE FOURTH GRADE TO HIGH SCHOOL GRADUATION CIRCLED ROUND AND ABOUT THE SOUTH.
ELEVEN YEARS, FOURTEEN DIFFERENT PUBLIC SCHOOLS.
NOMADIC EXISTENCE MADE NATURE MY COMPANION OF CHOICE.
THE OUTDOORS WAS THE ONE PART OF MY WORLD I PERCEIVED TO HOLD ROCK STEADY. ANIMALS AND PLANTS I COULD COUNT ON—HUMAN RELATIONSHIPS WERE MORE DIFFICULT.
WHAT SHOULD WE DO WITH THIS SPANISH MOSS?
YOU MEAN *TILLANDSIA USNEOIDES*, DON'T YOU?
OF COURSE. IT NEEDS TO BE BACK ON THOSE BRANCHES, TO DRAW NUTRIENTS AND MOISTURE FROM THE AIR.
I CREATED THREE BOYS—I, ME, AND MYSELF—IN MY HEAD.
I THOUGHT OF THEM AS FRIENDS, BUT THEY WERE REALLY JUST SELF-PITY.

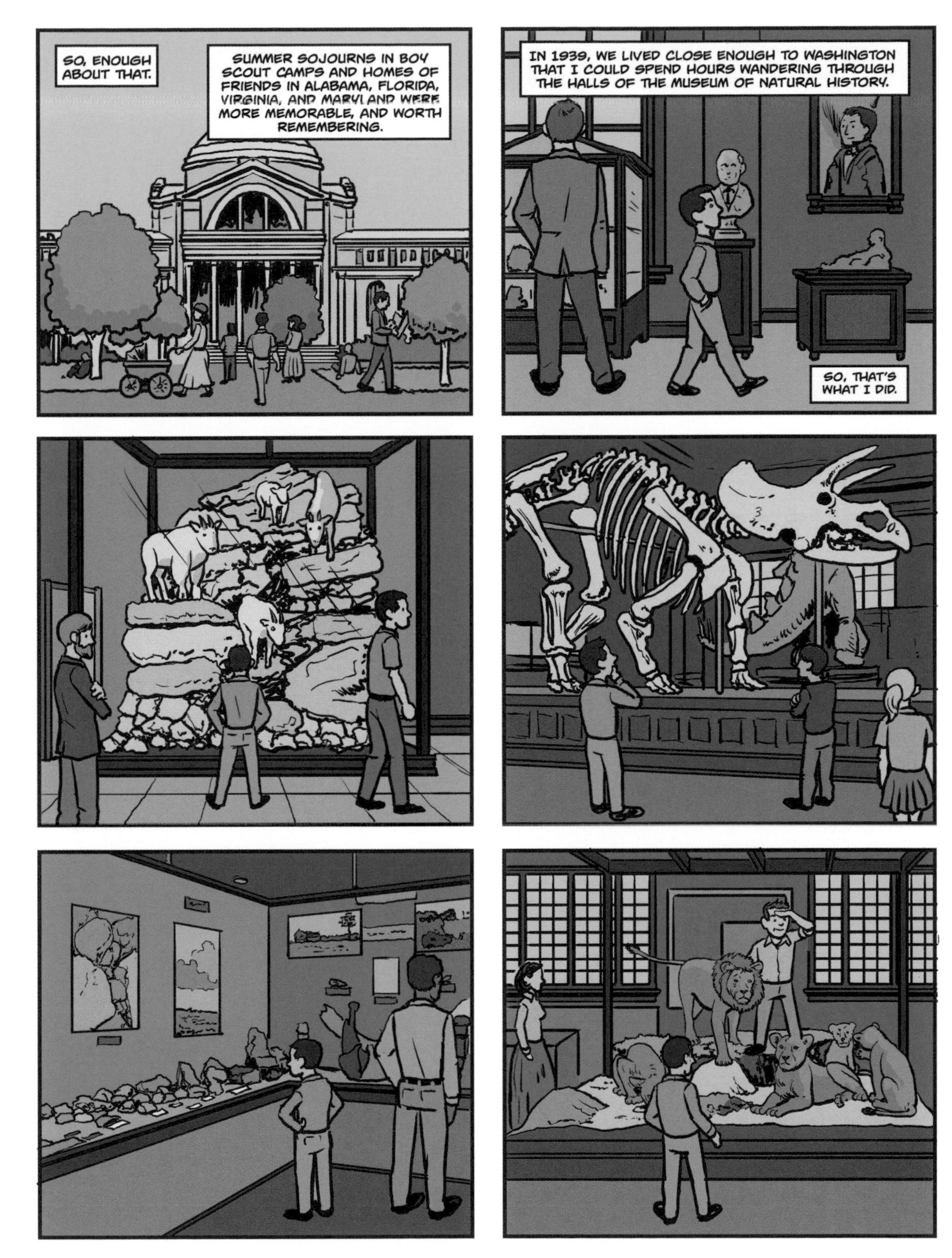

SO, ENOUGH ABOUT THAT.
SUMMER SOJOURNS IN BOY SCOUT CAMPS AND HOMES OF FRIENDS IN ALABAMA, FLORIDA, VIRGINIA, AND MARYLAND WERE MORE MEMORABLE, AND WORTH REMEMBERING.
IN 1939, WE LIVED CLOSE ENOUGH TO WASHINGTON THAT I COULD SPEND HOURS WANDERING THROUGH THE HALLS OF THE MUSEUM OF NATURAL HISTORY.
SO, THAT'S WHAT I DID.

BEYOND DREAMING OF DISTANT JUNGLES AND SAVANNAS, A NEW VISION OF SCIENTIFIC PROFESSIONALISM TOOK FORM.
BEHIND THE CLOSED DOORS ABOVE LABORED *THE CURATORS*, SHAMANS OF MY NEW WORLD.
I NEVER MET ONE OF THEM—PERHAPS A FEW PASSED ME UNRECOGNIZED IN THE EXHIBITION HALLS.

JUST THE AWARENESS OF THEIR EXISTENCE—EXPERTS OF SUCH HIGH ORDER DOING THE BUSINESS OF THE GOVERNMENT IN SPLENDID SURROUNDINGS—FIXED IN ME THE CONCEPTION OF SCIENCE AS A DESIRABLE LIFE GOAL.

I COULD NOT IMAGINE ANY ACTIVITY MORE ELEVATING THAN TO ACQUIRE THEIR KIND OF KNOWLEDGE, TO BE A STEWARD OF ANIMALS AND PLANTS, AND TO PUT THE EXPERTISE TO PUBLIC SERVICE.

SOUTHERN MAGNOLIAS, JAPANESE THREADLEAF MAPLE, AMERICAN ELM, WHITE OAK, ATLAS CEDAR...
DON'T FORGET SUGAR MAPLE AND NORTHERN RED OAK.
IT WAS A SHORT ONE-HOUR WALK TO THE SECOND FOCUS OF MY LIFE.
EXPLORING EVERY CAGE AND GLASS-WALLED ENCLOSURE.
STARING AT THE CHARISMATIC BIG ANIMALS.

THE NATIONAL ZOO, ALSO ADMINISTERED BY THE SMITHSONIAN, WAS A LIVING MUSEUM OF EQUAL POTENCY TO THE NATIONAL MUSEUM OF NATURAL HISTORY.
ELEPHANT
TIGERS
MONKEYS
INSECT HOUSE
SNAKES
RHINOS
REPTILES

RHINOCEROTIDAE

OPHIOPHAGUS HANNAH

CASUARIUS

HI.
HI.

HOW MANY BITES DO YOU THINK ONE OF THOSE WOULD NEED FOR...ONE OF US?
ONE OR TWO.
SNAP!

SPWHZM-MLASH!

TWO, FOR CERTAIN.

FRIEND?

SMALL ANIMAL ROOM
SEE YOU.
YUP. SEE YOU.
THERE WERE ALSO SMALLER ANIMALS THAT EVENTUALLY BECAME EQUALLY FASCINATING. I DEVELOPED A LIKING FOR LIZARDS, MARMOSETS, PARROTS, AND PHILIPPINE TREE RATS.

CLOSE TO THE ZOO WAS A THIRD SPECIAL PLACE–ROCK CREEK PARK, A WOODED URBAN RETREAT INTO WHICH I VENTURED ON "EXPEDITIONS."

IT HAD NO ELEPHANTS TO PHOTOGRAPH NOR TIGERS TO DROP-NET. BUT INSECTS WERE EVERYWHERE.
DO YOU UNDERSTAND THAT PART?
NOPE. I THINK IT'S REAL SCIENCE.
PRINCIPLES OF INSECT MORPHOLOGY

ON EXCURSIONS THERE WITH MY NEW BEST FRIEND, ELLIS MACLEOD, I ACQUIRED A PASSION FOR BUTTERFLIES.

VANESSA ATALANTA
SPEYERIA CYBELE

WE ALSO SOUGHT THE ELUSIVE MOURNING CLOAK–NYMPHALIS ANTIOPA–ALONG THE SHADED TRAILS OF ROCK CREEK.
RATS!

ROCK CREEK PARK BECAME UGANDA AND SUMATRA WRIT SMALL, AND THE COLLECTION OF INSECTS I BEGAN TO ACCUMULATE AT HOME A SIMULACRUM OF THE NATIONAL MUSEUM.

AT ABOUT THIS TIME, I ALSO BECAME FASCINATED WITH ANTS.
ACANTHOMYOPS

DO YOU SMELL LEMONS? WOW!

THIRTY YEARS LATER, IN MY LABORATORY AT HARVARD, I DISCOVERED THAT CITRONELLA IS SECRETED BY GLANDS ATTACHED TO THE MANDIBLES OF THE ANTS.

LIKE THE PYGIDIAL SUBSTANCES OF THE PENSACOLA DORYMYRMEX, THEY USE IT TO ATTACK ENEMIES AND SPREAD ALARM THROUGH THE COLONY.

THAT DAY IN ROCK CREEK, THE LITTLE ARMY QUICKLY THINNED AND VANISHED INTO THE DARK INTERIOR OF THE STUMP HEARTWOOD.
THAT SMELL IS SOMETHING ELSE.
YEAH, IT'S PRETTY GREAT. I WONDER WHAT IT'S FOR? AND...

WHAT DO YOU THINK THEY'RE DOING UNDER THERE?

FIELDBOOK OF INSECTS

ELLIS BECAME A PROFESSOR OF ENTOMOLOGY, SO, WE BOTH ENDED UP DOING REAL SCIENCE...
NOT TOGETHER, BUT WE REMAINED GOOD FRIENDS ALL OUR LIVES, LONG AFTER MY FATHER'S MIGRATION PULLED US APART.
WE ARRIVED IN MOBILE IN 1941, NOT LONG AFTER MY GRANDMOTHER DIED, TO LIVE IN THE HOUSE MY FATHER AND HIS BROTHER INHERITED.
MOST OF THE MEN IN THE WILSON CLAN HAD EITHER DIED OR DISPERSED, LEAVING BEHIND WIDOWS AND UNMARRIED DAUGHTERS.
MY FATHER, AT THIS POINT, TURNED OUT TO BE A FAMILY HISTORIAN, SEIZED BY NOSTALGIA AND A YEARNING FOR REFLECTED GLORY.
THE REMINISCING DRONED ON, A RECYCLING OF THE STORIES AND SKETCHES OF OLD MOBILE.
GRANDMA MAY WOULD ALWAYS...
SO AUNT HOPE...
...BUT AUNT GEORGIA,
...ANYWAY, AUNT SARAH
WAR BETWEEN THE STATES
OCCASIONALLY, WE VISITED MAGNOLIA CEMETERY, WHERE OUR FOREBEARS AND THEIR MULTITUDINOUS RELATIONS AND FRIENDS LAY AT REST.
DO YOU THINK HE'LL NOTICE IF I...
I DON'T THINK SO.
GO AHEAD. I'LL TELL HIM I SAID IT WAS OKAY.
PEARL AND I STOOD PATIENTLY BY AS MY FATHER LOCATED GRAVES, CHECKED DATES, AND RECONSTRUCTED LIVES AND GENEALOGIES.

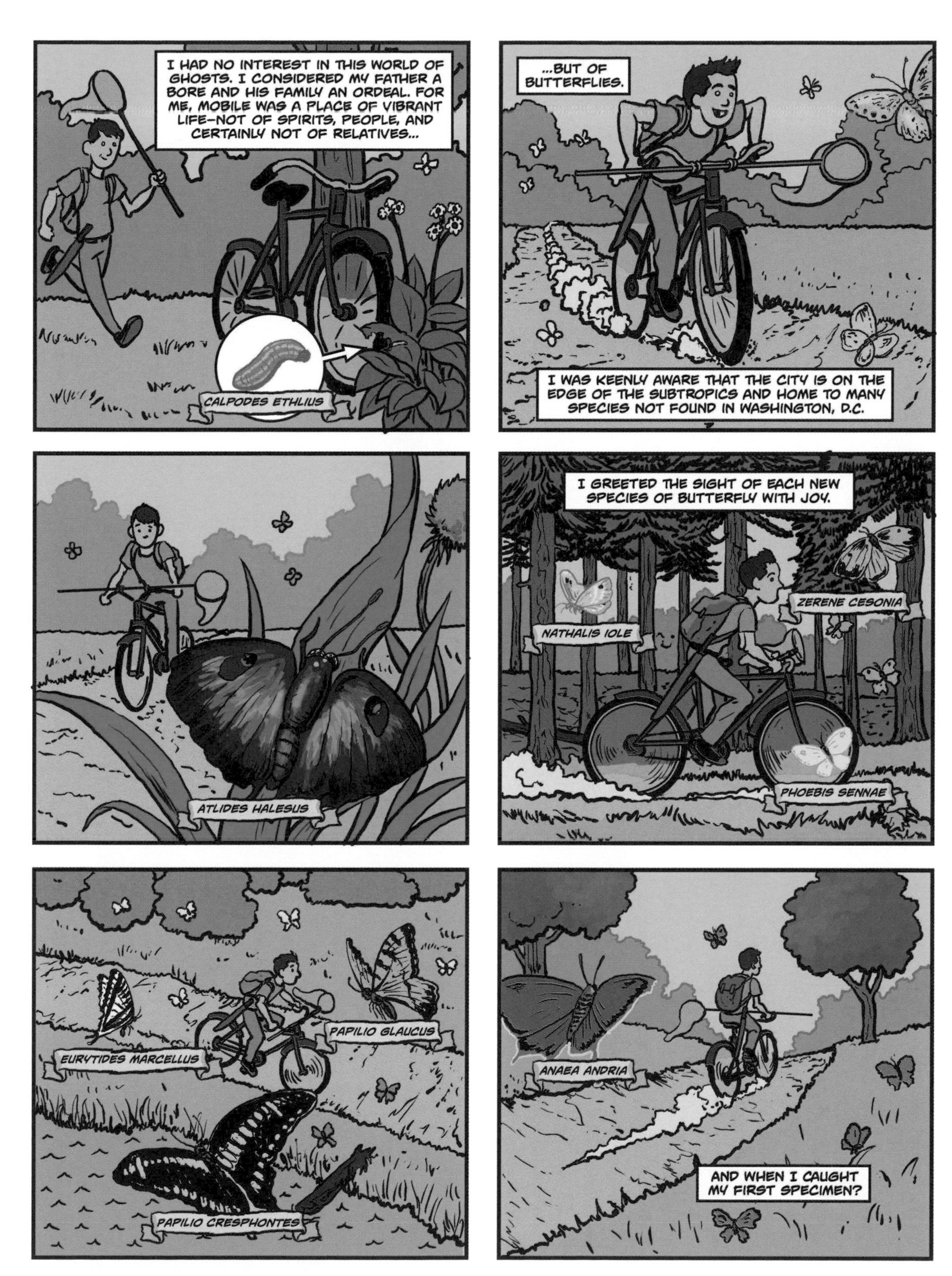

I HAD NO INTEREST IN THIS WORLD OF GHOSTS. I CONSIDERED MY FATHER A BORE AND HIS FAMILY AN ORDEAL. FOR ME, MOBILE WAS A PLACE OF VIBRANT LIFE—NOT OF SPIRITS, PEOPLE, AND CERTAINLY NOT OF RELATIVES...
CALPODES ETHLIUS
...BUT OF BUTTERFLIES.
I WAS KEENLY AWARE THAT THE CITY IS ON THE EDGE OF THE SUBTROPICS AND HOME TO MANY SPECIES NOT FOUND IN WASHINGTON, D.C.
ATLIDES HALESUS
I GREETED THE SIGHT OF EACH NEW SPECIES OF BUTTERFLY WITH JOY.
ZERENE CESONIA
NATHALIS IOLE
PHOEBIS SENNAE
PAPILIO GLAUCUS
EURYTIDES MARCELLUS
PAPILIO CRESPHONTES
ANAEA ANDRIA
AND WHEN I CAUGHT MY FIRST SPECIMEN?

I CONSIDERED MYSELF A BIG-GAME HUNTER.

DURING THE NEXT TWO YEARS, BEFORE WE HIT THE ROAD AGAIN, AS IT SEEMED INEVITABLY WE MUST, MY INTEREST IN NATURAL HISTORY SOARED.
I WENT LOOKING FOR PILEATED WOODPECKERS RUMORED TO NEST AT SPANISH FORT, AND ON THE WAY, SAW MY FIRST WILD ALLIGATORS.
I BUILT A SECRET OUTDOOR SHELTER PARTLY FROM THE STEMS OF POISON OAK...
...AND PAID FOR IT.
RHUS QUERCIFOLIA, A HUNDRED PACES, DEAD AHEAD.
AND I LEARNED THE CORRECT MANEUVER FOR CATCHING CAROLINA ANOLE LIZARDS.

IT DIDN'T ALWAYS WORK, OF COURSE, BUT IT DID MOST OF THE TIME.
I ALSO GOT GOOD AT CATCHING SNAKES.

I LIKED REPTILES.
EDWARD OSBORNE WILSON!
WHAT ARE YOU DOING?

CAUGHT A COACHWHIP SNAKE.
LOOK HOW LONG IT IS!

AHHH...
YES, I SEE. VERY LONG.
IT IS NOW 8:10 PM. SUNSET IS WHEN?
MY ALMANAC IS IN THE OTHER ROOM. WOULD YOU HOLD THIS WHILE I LOOK IT UP, PLEASE?
I DON'T THINK SO. JUST GIVE ME YOUR BEST ESTIMATE, EDWARD.
UM. 9:30?
VERY GOOD. SO. YOU WILL WALK 40 MINUTES IN WHATEVER DIRECTION YOU CHOOSE, AT WHICH POINT YOU WILL RELEASE THAT SNAKE. THEN YOU WILL WALK BACK.
EMPTY-HANDED.
NOW MARCH. I'LL PUT THESE IN YOUR ROOM.
YES MA'AM.

IN JUNE 1942 ELLIS MACLEOD CAME DOWN FROM WASHINGTON TO STAY WITH ME FOR THE SUMMER.
WAIT'LL YOU SEE WHAT WE HAVE DOWN HERE.

WE VISITED MY FAVORITE HAUNTS...

SHARED AGAIN OUR OLD FANTASIES...

AND WE RENEWED OUR INTENTION TO BECOME ENTOMOLOGISTS.

AFTER ELLIS RETURNED HOME THAT FALL, I SET OUT TO COLLECT AND STUDY ALL THE ANTS IN A VACANT LOT NEXT TO THE CHARLESTON STREET HOUSE.

I STILL REMEMBER THE SPECIES I FOUND, IN VIVID DETAIL...
A COLONY OF TRAP-JAWED ODONTOMACHUS HAEMATODUS WHOSE VICIOUS STINGS DROVE ME AWAY FROM THEIR NEST.
PHEIDOLE, POSSIBLY PHEIDOLE FLORIDANUS.
AND A COLONY OF FIRE ANTS.
SOLENOPSIS FIRE ANTS, A NEW KIND, UNMISTAKABLY!
THEIR DISCOVERY WAS THE EARLIEST RECORD OF THE SPECIES IN THE UNITED STATES, AND I LATER PUBLISHED IT AS A DATUM IN A TECHNICAL ARTICLE.
YOUR FIRST SCIENTIFIC OBSERVATION.

MY ENERGIES AND CONFIDENCE WERE GATHERING. BY THE FALL OF 1942, AT THE AGE OF THIRTEEN, I HAD BECOME IN EFFECT A CHILD WORKAHOLIC.

SOON AFTER THE START OF WWII THERE WAS A SHORTAGE OF CARRIERS FOR THE MOBILE PRESS REGISTER, AND I TOOK A JOB WITH BACKBREAKING HOURS, WITHOUT ADULT COERCION OR EVEN ENCOURAGEMENT.

SOMEHOW, FOR REASONS I DO NOT RECALL, AN ADULT DELIVERY SUPERVISOR LET ME TAKE OVER A MONSTER ROUTE–420 PAPERS IN THE CENTRAL CITY AREA.
THE RESIDENCES RECEIVING THE PAPERS WERE NOT WIDELY SPACED SUBURBAN HOUSES BUT CITY DWELLINGS, APARTMENT BUILDINGS WITH TWO OR THREE STORIES.

I COULD DELIVER ABOUT TWO PAPERS A MINUTE, MAKE TWO ROUND TRIPS BACK TO THE DOCK, AND BE HOME IN TIME TO GET BREAKFAST AND LEAVE FOR SCHOOL BY 8:00.

PRESS~REGISTER
I MADE $13 A WEEK, FROM WHICH I BOUGHT PARTS FOR MY BIKE AND WHATEVER CANDY, SOFT DRINKS, AND MOVIE TICKETS I WANTED
AND, MOST IMPORTANTLY, BOY SCOUT PARAPHERNALIA.

THE BOY SCOUTS OF AMERICA SEEMED INVENTED JUST FOR ME.
HANDBOOK FOR BOYS
BOY SCOUTS OF AMERICA
50¢
RULES, UNIFORMS, AND A CRYSTAL-CLEAR SET OF PRACTICAL ETHICS TO LIVE BY.

EVEN TODAY, IF I PHYSICALLY JOG MY MEMORY, I CAN STILL RECITE THE OATH...
ON MY HONOR I WILL DO MY BEST: TO DO MY DUTY...

...TO GOD AND MY COUNTRY, AND TO OBEY THE SCOUT LAW. TO HELP OTHER PEOPLE AT ALL TIMES...

...TO KEEP MYSELF PHYSICALLY STRONG, MENTALLY AWAKE AND MORALLY STRAIGHT.
I DRANK IN AND ACCEPTED EVERY WORD. STILL DO, AS RIDICULOUS AS THAT OATH MAY SEEM TO MY COLLEAGUES IN THE INTELLECTUAL TRADE...

...TO WHOM I SAY... "LET'S SEE YOU DO BETTER IN FIFTY-FOUR WORDS OR LESS."
FOR MYSELF, I HAPPILY CRUNCHED THROUGH PROGRAMS FOR SUBJECTS AS DIVERSE AS BIRD STUDY, FARM RECORDS AND BOOK-KEEPING, LIFE SAVING, JOURNALISM, AND PUBLIC HEALTH.
AND MY HEART SANG WHEN I FIRST READ THE PRESCRIPTION FOR INSECT LIFE.
TO OBTAIN THIS MERIT BADGE, A SCOUT MUST GO INTO THE COUNTRY WITH THE EXAMINER AND...
...SHOW TO HIM THE NATURAL SURROUNDINGS IN WHICH CERTAIN SPECIFIED INSECTS LIVE...
...FIND AND DEMONSTRATE LIVING SPECIMENS OF THE INSECTS
PHEIDOLE FLORIDAN
COMMONLY KNOWN A
OKAY ED, WE'RE ALL SET.
...AND TELL OF THEIR HABITS OR OF THE NATURE OF THEIR FITNESS FOR LIFE IN THEIR PARTICULAR SURROUNDINGS.
ED. ED! I HAVE
WHAT I
BUT WAIT, SIR. WE HAVEN'T EVEN GOTTEN TO BUTTERFLIES. OR LACEWINGS.
MY FRIEND ELLIS AND I STUDIED
NO, REALLY, ED.
I'M SATISFIED.

SCOUTING PROVED TO BE THE IDEAL SOCIALIZING ENVIRONMENT FOR AN UNDERSIZED, INTROVERTED ONLY CHILD.
HEY BUGGY WHAT'S THIS THINGEE OVER HERE? DANGEROUS?

IT ALSO ADDED ANOTHER DIMENSION TO MY EXPANDING NICHE.
IN THE SUMMER OF 1943 I WAS ASKED TO BE THE NATURE COUNSELOR AT CAMP PUSHMATAHA.

AT FOURTEEN, I WAS THE YOUNGEST COUNSELOR WITH NO EXPERIENCE AT INSTRUCTION, BUT I QUICKLY FIGURED OUT WHAT INTERESTED OTHER BOYS...

WHAT WOULD GET THEM TALKING ABOUT NATURAL HISTORY AND MAKE THEM RESPECT THE SUBJECT.

HISS!

SNAKES.

SO I BECAME A TEACHER.
SNAKE!
SNAKE!
AND OFF I'D GO TO PERFORM MY DERRING-DO, WHICH I FOLLOWED WITH A BRIEF LECTURE ON THE SPECIES DISCOVERED.
THE PYGMY RATTLESNAKE–*SISTRURUS MILIARIUS*–GROWS TO NO MORE THAN FIFTY CENTIMETERS IN LENGTH
THEY ARE LESS DEADLY THAN THEIR LARGER COUINS, THE DIAMONDBACK AND CANEBRAKE RATTLESNAKES.
THEY ARE STILL POISONOUS AND MODERATELY...

#$%^!

OFF I WENT WITH AN ADULT COUNSELOR TO A NEARBY DOCTOR, WHO ADMINISTERED THE OLD FASHIONED FIRST-AID TREATMENT AS QUICKLY AS HE COULD.
I KNEW THE DRILL– I HAD LEARNED IT FOR THE REPTILE LIFE MERIT BADGE.

I DIDN'T CRY, BUT I DID CURSE LOUDLY AND NONSTOP AT MYSELF FOR MY STUPIDITY.
$#!
%#$!
%?#!
I KNEW A GREAT DEAL OF OFF-COLOR LANGUAGE AT FOURTEEN, WHICH MUST HAVE SURPRISED THE ADULTS.

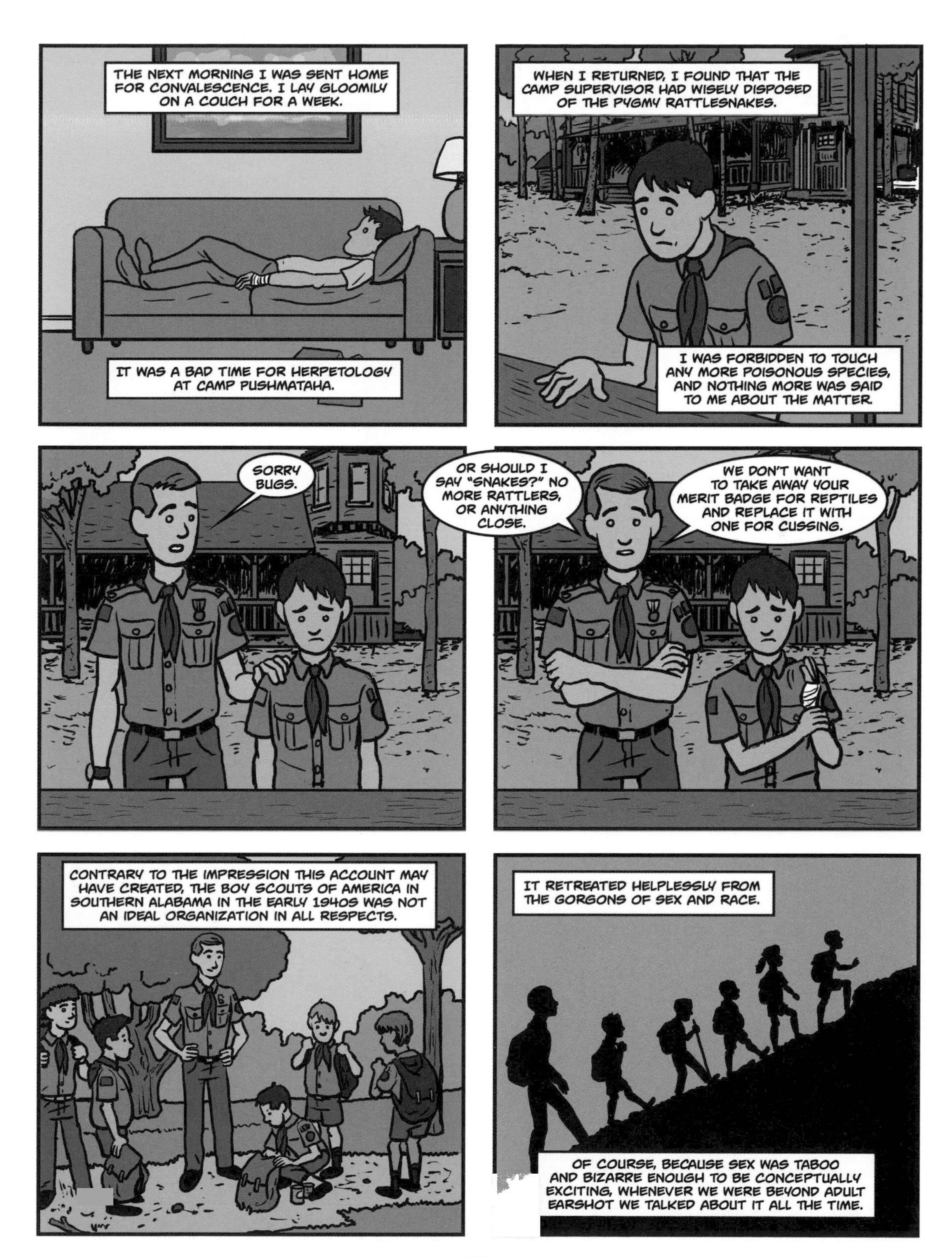
THE NEXT MORNING I WAS SENT HOME FOR CONVALESCENCE. I LAY GLOOMILY ON A COUCH FOR A WEEK.
IT WAS A BAD TIME FOR HERPETOLOGY AT CAMP PUSHMATAHA.
WHEN I RETURNED, I FOUND THAT THE CAMP SUPERVISOR HAD WISELY DISPOSED OF THE PYGMY RATTLESNAKES.
I WAS FORBIDDEN TO TOUCH ANY MORE POISONOUS SPECIES, AND NOTHING MORE WAS SAID TO ME ABOUT THE MATTER.
SORRY BUGS.
OR SHOULD I SAY "SNAKES?" NO MORE RATTLERS, OR ANYTHING CLOSE.
WE DON'T WANT TO TAKE AWAY YOUR MERIT BADGE FOR REPTILES AND REPLACE IT WITH ONE FOR CUSSING.
CONTRARY TO THE IMPRESSION THIS ACCOUNT MAY HAVE CREATED, THE BOY SCOUTS OF AMERICA IN SOUTHERN ALABAMA IN THE EARLY 1940S WAS NOT AN IDEAL ORGANIZATION IN ALL RESPECTS.
IT RETREATED HELPLESSLY FROM THE GORGONS OF SEX AND RACE.
OF COURSE, BECAUSE SEX WAS TABOO AND BIZARRE ENOUGH TO BE CONCEPTUALLY EXCITING, WHENEVER WE WERE BEYOND ADULT EARSHOT WE TALKED ABOUT IT ALL THE TIME.

THAT'S WHERE I LEARNED ALL MY FILTHY LANGUAGE —FROM OTHER BOYS.
NO WAY.
I SWEAR, MY COUSIN KNOWS A GUY WHO
HA!HA!HA!HA!HA!HA!H
WE APPROACHED SEX OBLIQUELY, WITH RAUCOUS JOKES THAT TOUCHED ON EVERY IMAGINABLE GROTESQUERIE.
BUT DID YOU EVER, YOU KNOW, KISS HER?
BUT OF TYPICAL HETEROSEXUAL RELATIONS? SCANT WAS KNOWN AND NOTHING SAID.
RACE WAS ALSO EMPHATICALLY NOT ON THE OFFICIAL AGENDA. ALL THE BOYS IN THE HANDBOOK FOR BOYS WERE WHITE, AND SO WERE ALL THE SCOUTS I KNEW.
ED, MAY I INTERRUPT FOR A MOMENT?
NEW TROOPS FORMED ALL THE TIME, THOUGH, AND I WAS INVITED TO GIVE A SHORT TALK TO ONE JUST STARTING UP IN A BLACK RURAL AREA NEAR BREWTON, ALABAMA.
I DID NOT FEEL PRIDE IN THE EXAMPLE I WAS SUPPOSED TO HAVE SET.
...TO HELP OTHER PEOPLE AT ALL TIMES, TO KEEP MYSELF PHYSICALLY STRONG, MENTALLY AWAKE AND MORALLY STRAIGHT.
I FELT SHAME, AND WAS DEPRESSED FOR DAYS. I KNEW IN MY HEART THAT THOSE BOYS WOULD HAVE FEW REAL ADVANTAGES.
NO MATTER HOW GIFTED THEY WERE OR HOW HARD THEY TRIED, THE DOORS OPEN TO ME WERE SHUT TO THEM.

THEN I GRADUALLY FORGOT ABOUT THE MATTER. WHAT COULD I DO?
MY MIND WAS ON OTHER THINGS. I WAS FILLED WITH AMBITION AND ANXIETY AND DID NOT HAVE A STRONG SOCIAL CONSCIENCE.
CIVIL RIGHTS ACTIVISTS WHO RISKED THEIR LIVES TO BREAK SEGREGATION WERE HEROES TO MY LIKING—SINGLE MINDEDLY TRUE TO A MORAL CODE, PHYSICALLY COURAGEOUS, ENDURING.
THEY MADE ME LOOK AGAIN AT MY SOCIAL HERITAGE.
BUT I CANNOT CLAIM TO HAVE BEEN A LIBERAL AS A BOY AND YOUNG MAN.
CERTAINLY NOT ONE WITH ANY FORESIGHT OR COURAGE.

IN THE FALL OF 1943, I CAME BACK TO SPEND ANOTHER YEAR WITH BELLE RAUB.
AT FOURTEEN I WAS OLD ENOUGH TO BE BAPTIZED AND BORN AGAIN BY MY OWN FREE WILL. NO ONE COUNSELED ME TO TAKE THIS STEP.

ONE EVENING, WHILE LISTENING TO A RECITAL, IT JUST HAPPENED.
WERE YOU THERE WHEN THEY CRUCIFIED MY LORD? WERE YOU THERE WHEN THEY CRUCIFIED MY LORD? SOMETIMES IT CAUSES ME TO TREMBLE, TREMBLE.

WERE YOU THERE WHEN THEY CRUCIFIED MY LORD?
I FELT EMOTION AS THOUGH FROM THE LOSS OF A FATHER, BUT ONE RETRIEVABLE BY REDEMPTION. I WANTED TO DO SOMETHING DECISIVE.
A CIGAR!
A CIGAR!

SO MOTHER RAUB AND I CALLED ON REVEREND WALLACE ROGERS AT THE FIRST BAPTIST CHURCH TO ANNOUNCE MY DECISION.

CONGRATULATIONS, SON. WHEN DO YOU WANT TO DO THIS?
MOTHER RAUB SAID NOTHING, THEN OR LATER, ABOUT HIS TRANSGRESSION. BUT I KNEW WHAT WAS ON HER MIND!

IN HIS FRIENDLY, CASUAL WAY HE CONGRATULATED ME ON MY DECISION, AND TOGETHER WE CHOSE A DATE FOR THE BAPTISM.

ONE SUNDAY EVENING IN FEBRUARY 1944, ROGERS RECITED THE BAPTISMAL DEDICATION...

THAT'S IT?
IT WAS KIND OF...PHYSICAL.
GASP!

LIKE...GOING SWIMMING IN PENSACOLA BAY.
OR SOMETHING

I HAD FELT EMBARRASSED AND UNCOMFORTABLE DURING THE BAPTISM. WAS THE WHOLE WORLD COMPLETELY PHYSICAL, AFTER ALL?
AND WHAT ABOUT DR. ROGERS' CLOTHING, AND CIGAR?

AND WITH THOSE THOUGHTS SOMETHING SMALL, SOMEWHERE, CRACKED.

I HAD BEEN HOLDING AN EXQUISITE, PERFECT SPHERICAL JEWEL IN MY HAND, AND NOW...

...TURNING IT OVER IN A CERTAIN LIGHT, I DISCOVERED A RUINOUS FRACTURE.

THE FAITHFUL MIGHT SAY I NEVER TRULY KNEW GRACE, NEVER HAD IT.
BUT THEY'D BE WRONG. THE TRUTH IS THAT I FOUND IT AND ABANDONED IT. IN THE YEARS FOLLOWING, I DRIFTED AWAY FROM THE CHURCH.

SCIENCE BECAME THE NEW LIGHT AND THE WAY.

BUT WHAT OF RELIGION? THERE MUST BE AN EXPLANATION FOR IT, FOR MORAL PRECEPTS AND THE RITES OF PASSAGE AND THE CRAVING FOR IMMORTALITY.

ATOMS TO GENES TO THE HUMAN SPIRIT... RELIGION HAS TO BE EXPLAINED AS A MATERIAL PROCESS, FROM THE BOTTOM UP.

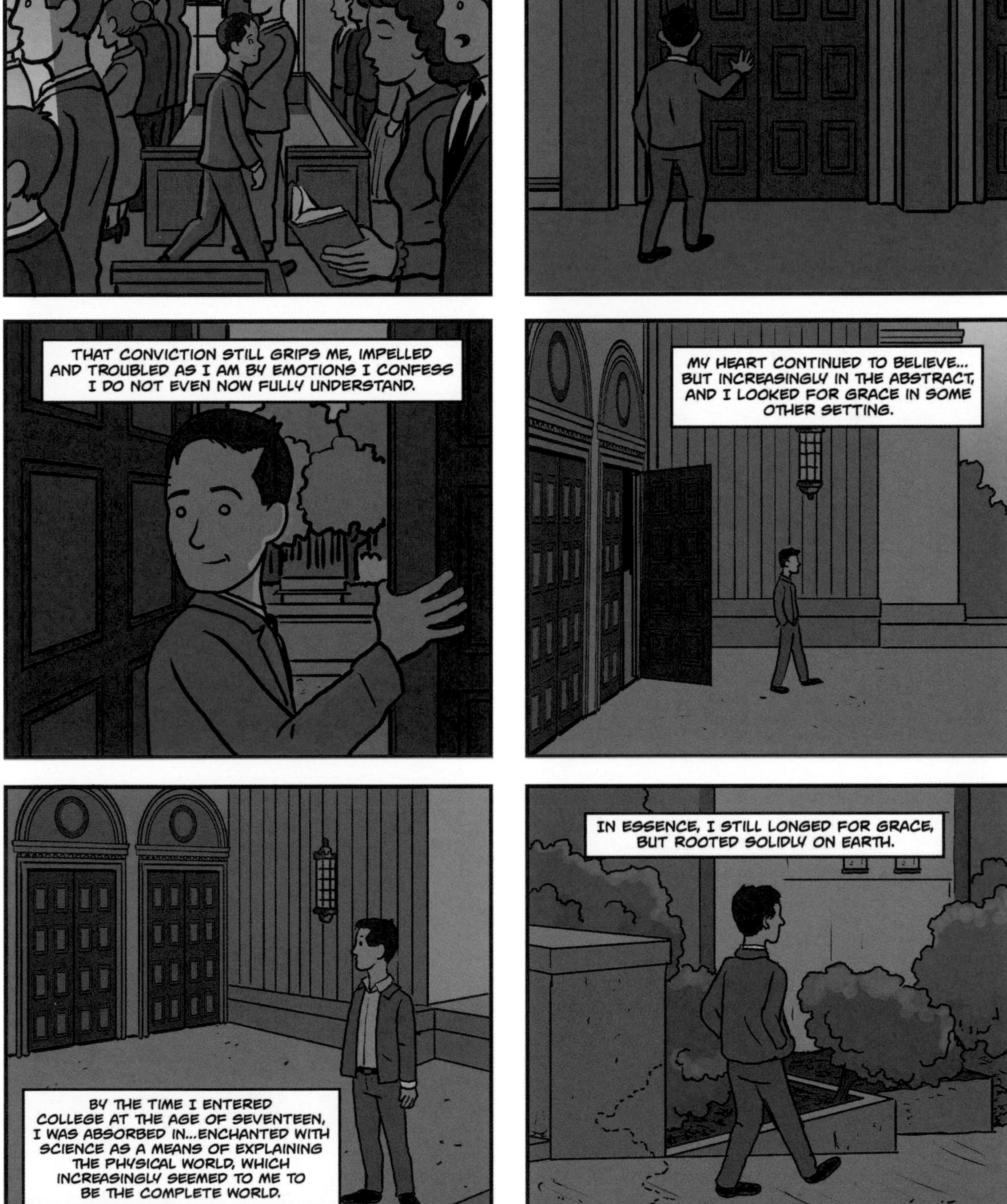
THAT CONVICTION STILL GRIPS ME, IMPELLED AND TROUBLED AS I AM BY EMOTIONS I CONFESS I DO NOT EVEN NOW FULLY UNDERSTAND.
MY HEART CONTINUED TO BELIEVE... BUT INCREASINGLY IN THE ABSTRACT, AND I LOOKED FOR GRACE IN SOME OTHER SETTING.
BY THE TIME I ENTERED COLLEGE AT THE AGE OF SEVENTEEN, I WAS ABSORBED IN...ENCHANTED WITH SCIENCE AS A MEANS OF EXPLAINING THE PHYSICAL WORLD, WHICH INCREASINGLY SEEMED TO ME TO BE THE COMPLETE WORLD.
IN ESSENCE, I STILL LONGED FOR GRACE, BUT ROOTED SOLIDLY ON EARTH.

THE UNIVERSITY OF ALABAMA SAVED ME.
BUT FIRST, I HAD TO GET THROUGH HIGH SCHOOL. AND HIGH SCHOOL FOOTBALL.
AT 112 POUNDS I WAS ALLOWED TO STRAP ON A UNIFORM BECAUSE THE TEAM NEEDED EVERY MAN (WELL, EVERY BOY) IT COULD GET.
FOOTBALL WAS WHAT YOUNG MALES ASPIRED TO DO WHEN NOT IN CLASS OR OCCUPIED WITH PART-TIME JOBS.
AND THE ONES WITH HEAVY SHOULDERS AND QUICK HANDS...THE ONES WITH NICKNAMES SUCH AS BUBBA, J. C., AND SKEETER?
THEY COULD HOPE FOR COLLEGE ATHLETIC SCHOLARSHIPS.
I WAS AT THE OTHER END OF THE STATISTICAL CURVE OF ATHLETIC PROMISE, AND MY NICKNAME WAS NOT IN THE GOOD 'OL BOY MOLD.
C'MON THERE, SNAKE. CONCENTRATE!
MY NICKNAME WAS BECAUSE OF MY BODY SHAPE, AND CERTAINLY NOT BECAUSE I COULD WEAVE THROUGH CROWDS OF TACKLERS HEAD DOWN WITH THE BALL TUCKED HARD ON MY WAIST, EXCEPT IN MY DREAMS.

IT WAS BECAUSE, WHILE A SWAMP FILLED WITH SNAKES MAY BE A NIGHTMARE TO MOST, FOR ME?
FOR ME IT WAS A CEASELESSLY ROTATING LATTICE OF WONDERS. EASTERN RIBBON SNAKES...
THAMNOPHIS SAURITUS SAURITUS
GREEN WATER SNAKES, LONG, HEAVY-BODIED, AND UNPLEASANT TO CATCH...
NERODIA CYCLOPION
MUD SNAKES, SPORTING A HARDENED TAIL TIP THEY USE TO HOLD DOWN THEIR PREY.
FARANCIA ABACURA
AMPHIUMA
AGKISTRODON PISCIVORUS
THE TIGERS AND LORDS OF THIS PLACE WERE THE POISONOUS COTTONMOUTH MOCCASINS, LARGE SEMIAQUATIC PIT VIPERS WITH THICK BODIES AND TRIANGULAR HEADS.

ONE DAY I MET AN OUTSIZED ADULT THAT MIGHT EASILY HAVE KILLED ME.

THIS SNAKE WAS NEARLY MY SIZE AS WELL AS VIOLENT AND NOISY— A COLLEAGUE, SO TO SPEAK.
I WAS THRILLED AT THE SIGHT, AND THE SNAKE LOOKED AS THOUGH IT COULD BE CAPTURED.

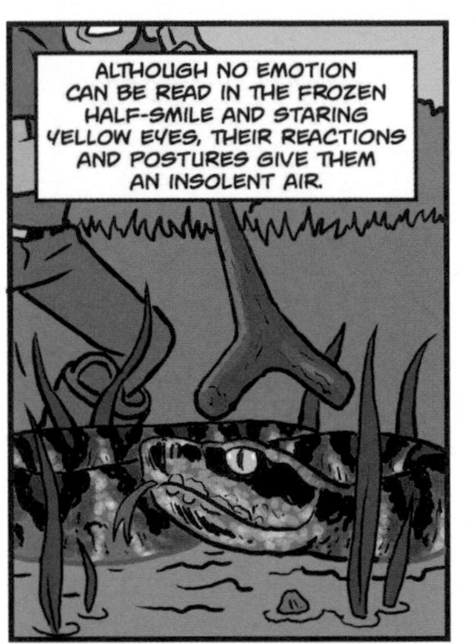
ALTHOUGH NO EMOTION CAN BE READ IN THE FROZEN HALF-SMILE AND STARING YELLOW EYES, THEIR REACTIONS AND POSTURES GIVE THEM AN INSOLENT AIR.

I MOVED INTO THE SNAKE HANDLER'S ROUTINE...

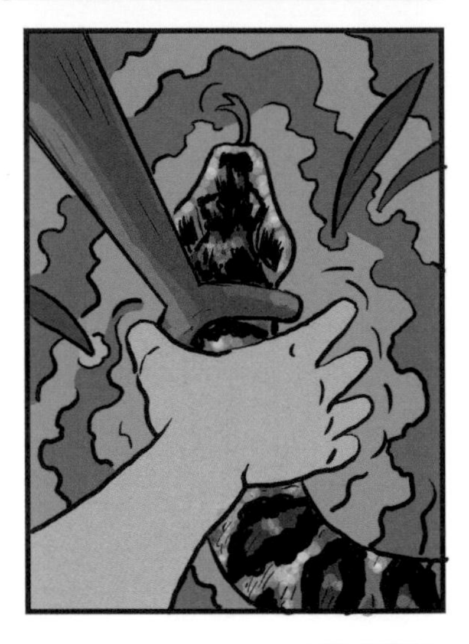

THE MORNING HEAT BECAME NOTICEABLE, REALITY CRASHED THROUGH, AND AT LAST I AWOKE FROM MY DREAM.
$&%! IF I GET BITTEN, WHO WOULD FIND ME?
WHAT AM I DOING IN THIS PLACE, ALONE?
PANT! PANT!
I STILL DON'T KNOW ALL THE REASONS.
IT CERTAINLY WASN'T FOR STATUS—I NEVER TOLD ANYONE MOST OF WHAT I DID.

MY FATHER AND PEARL SAW LITTLE VALUE IN MY SWAMP EXPEDITIONS, AND, LOOKING BACK, I CANNOT BLAME THEM.

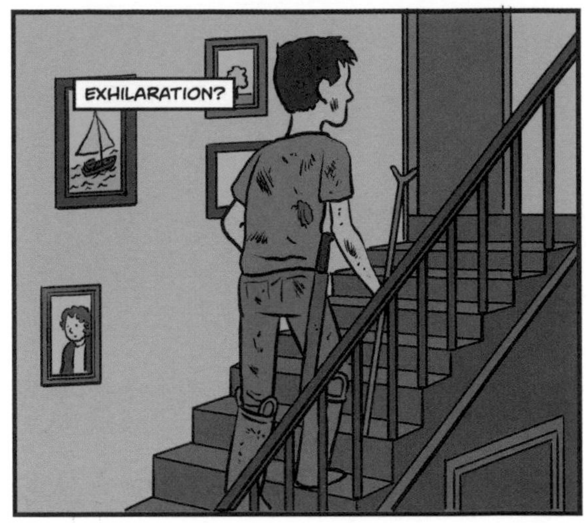
EXHILARATION?

POSSESSIVENESS?

VANITY?

AMBITION?

ALL OF THOSE. AND SOMETHING MORE.
AN UNDECIPHERED RESIDUE, A YEARNING REMAINING DEEP WITHIN ME THAT I HAVE NEVER UNDERSTOOD, NOR WISH TO FOR FEAR THAT IF NAMED IT MIGHT VANISH.

SUDDENLY, IN THE FALL OF 1945, SIXTEEN YEARS OLD AND COLLEGE ONLY A YEAR AWAY, I RECOGNIZED THAT I MUST GET SERIOUS ABOUT MY ENTOMOLOGY CAREER.
I HAVE TO DECIDE WHICH GROUP OF INSECTS I SHOULD BECOME A WORLD AUTHORITY ON.
AH.
YES, I SUPPOSE YOU DO.
WAR IN THE PACIFIC
BUTTERFLIES ARE OUT—TOO WELL KNOWN AND ALREADY BEING STUDIED BY LOTS OF GOOD SCIENTISTS.
HOW ABOUT FLIES?
REALLY?
HOUSEFLIES AND MOSQUITOES GIVE THEM A BAD NAME, BUT MOST SPECIES ARE LITTLE JEWELS IN NATURE'S CLOCKWORK. I LIKED THEIR CLEAN LOOKS, ACROBATICS, AND INSOUCIANT MANNER.
THEY'RE EVERYWHERE, IN DAZZLING VARIETY, AND THEY'RE IMPORTANT TO THE ENVIRONMENT!
EVERYWHERE.
I JUST NEED TO ORDER SOME EQUIPMENT. KILLING JARS, SCHMITT SPECIMEN BOXES. YOU KNOW, THE ESSENTIALS.
ESSENTIALS.

HOWEVER, EVEN THOUGH FLIES ARE EVERYWHERE, THE SPECIAL LONG BLACK INSECT PINS I NEEDED WERE NOT.
IT WAS THE WAR. I'D BEEN SANGUINE ABOUT IT—FRANKLIN DELANO ROOSEVELT HAD ALREADY FIXED JUST ABOUT EVERYTHING ELSE.

AND SINCE THE DEMOCRATS AND JOE LOUIS—BOTH OF WHOM ENJOYED MY ALLEGIANCE—HAD ALWAYS WON FOR AS FAR BACK AS I COULD REMEMBER, I KNEW THIS CRISIS WOULD ALSO WORK OUT ALL RIGHT.
CONFERENCE AT YALTA ENDS
Big Three at Yalta

IT HAD, BUT IN 1945, CZECHOSLOVAKIA WAS STILL A WAR ZONE AND ABOUT TO FALL UNDER SOVIET OCCUPATION. AND GUESS WHERE THOSE PINS WERE MADE?

I CAN'T BELIEVE I CAN'T GET THEM HERE. IT'S A CRISIS!
SO SORRY TO HEAR IT, ED.
I'LL HAVE TO PICK SOMETHING ELSE, RIGHT?

WITHOUT PAUSE, I CAST ABOUT FOR ANOTHER GROUP OF INSECTS IN WHICH TO INVEST MY ENERGIES, ONE THAT COULD BE PRESERVED IN SMALL BOTTLES OF ALCOHOL OBTAINED LOCALLY. I QUICKLY HIT UPON ANTS.
OF COURSE, ANTS.
OF COURSE.

MY OLD ACQUAINTANCES, SOME OF MY EARLIEST PASSIONS.

FROM A LOCAL DRUGSTORE, I PURCHASED DOZENS OF FIVE-DRAM PRESCRIPTION BOTTLES, THE OLD-FASHIONED GLASS ONES WITH METAL SCREW TOPS.

I MADE OBSERVATION NESTS, TOO, AND PREPARED TO LAUNCH MY CAREER AS A MYRMECOLOGIST.

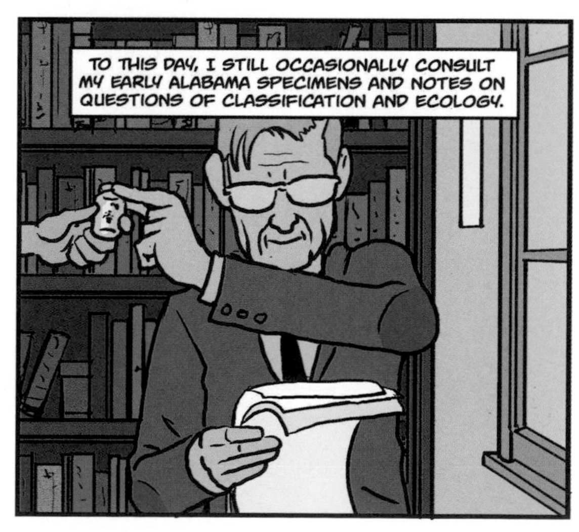
TO THIS DAY, I STILL OCCASIONALLY CONSULT MY EARLY ALABAMA SPECIMENS AND NOTES ON QUESTIONS OF CLASSIFICATION AND ECOLOGY.

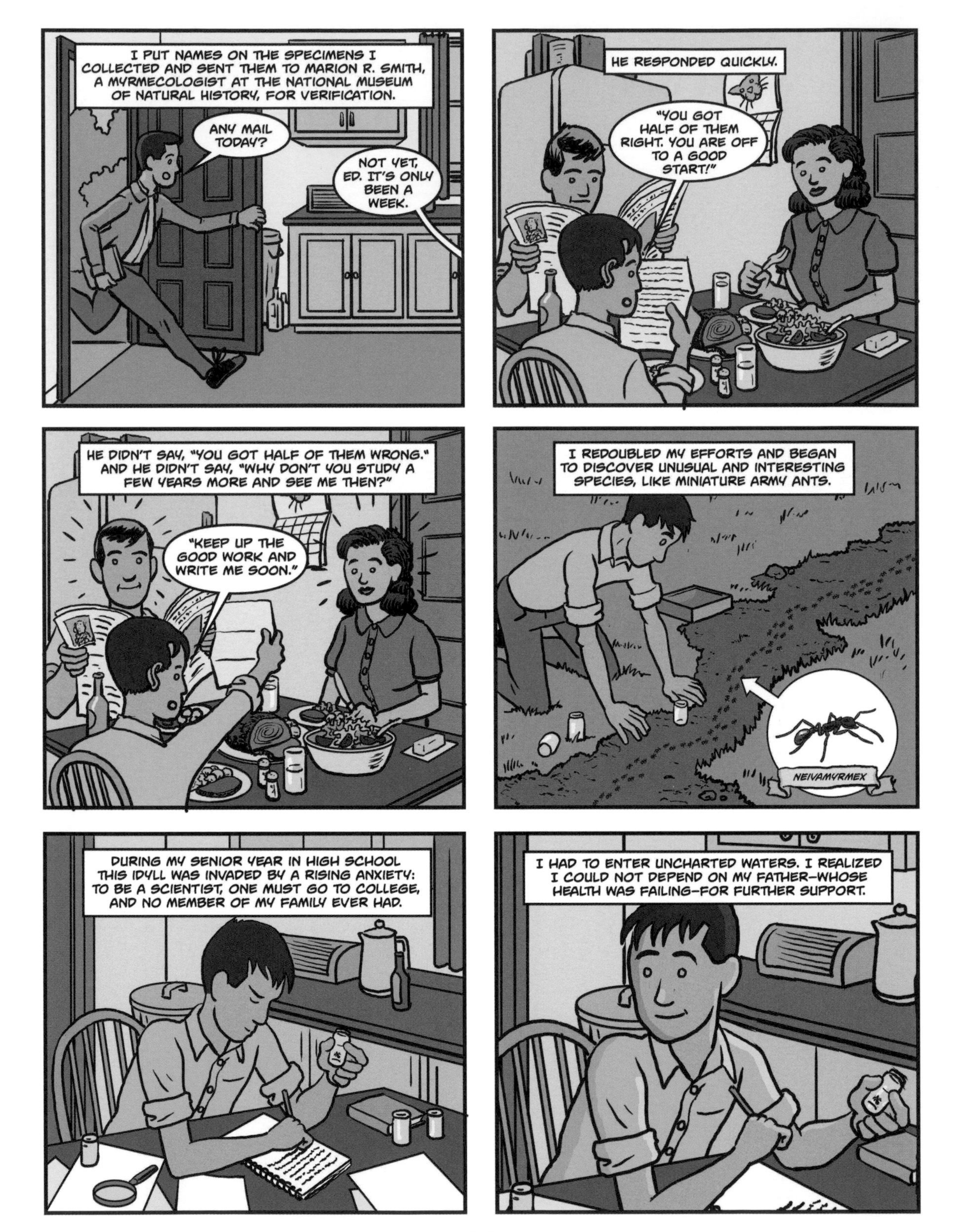
I PUT NAMES ON THE SPECIMENS I COLLECTED AND SENT THEM TO MARION R. SMITH, A MYRMECOLOGIST AT THE NATIONAL MUSEUM OF NATURAL HISTORY, FOR VERIFICATION.
ANY MAIL TODAY?
NOT YET, ED. IT'S ONLY BEEN A WEEK.
HE RESPONDED QUICKLY.
"YOU GOT HALF OF THEM RIGHT. YOU ARE OFF TO A GOOD START!"
HE DIDN'T SAY, "YOU GOT HALF OF THEM WRONG." AND HE DIDN'T SAY, "WHY DON'T YOU STUDY A FEW YEARS MORE AND SEE ME THEN?"
"KEEP UP THE GOOD WORK AND WRITE ME SOON."
I REDOUBLED MY EFFORTS AND BEGAN TO DISCOVER UNUSUAL AND INTERESTING SPECIES, LIKE MINIATURE ARMY ANTS.
NEIVAMYRMEX
DURING MY SENIOR YEAR IN HIGH SCHOOL THIS IDYLL WAS INVADED BY A RISING ANXIETY: TO BE A SCIENTIST, ONE MUST GO TO COLLEGE, AND NO MEMBER OF MY FAMILY EVER HAD.
I HAD TO ENTER UNCHARTED WATERS. I REALIZED I COULD NOT DEPEND ON MY FATHER—WHOSE HEALTH WAS FAILING—FOR FURTHER SUPPORT.

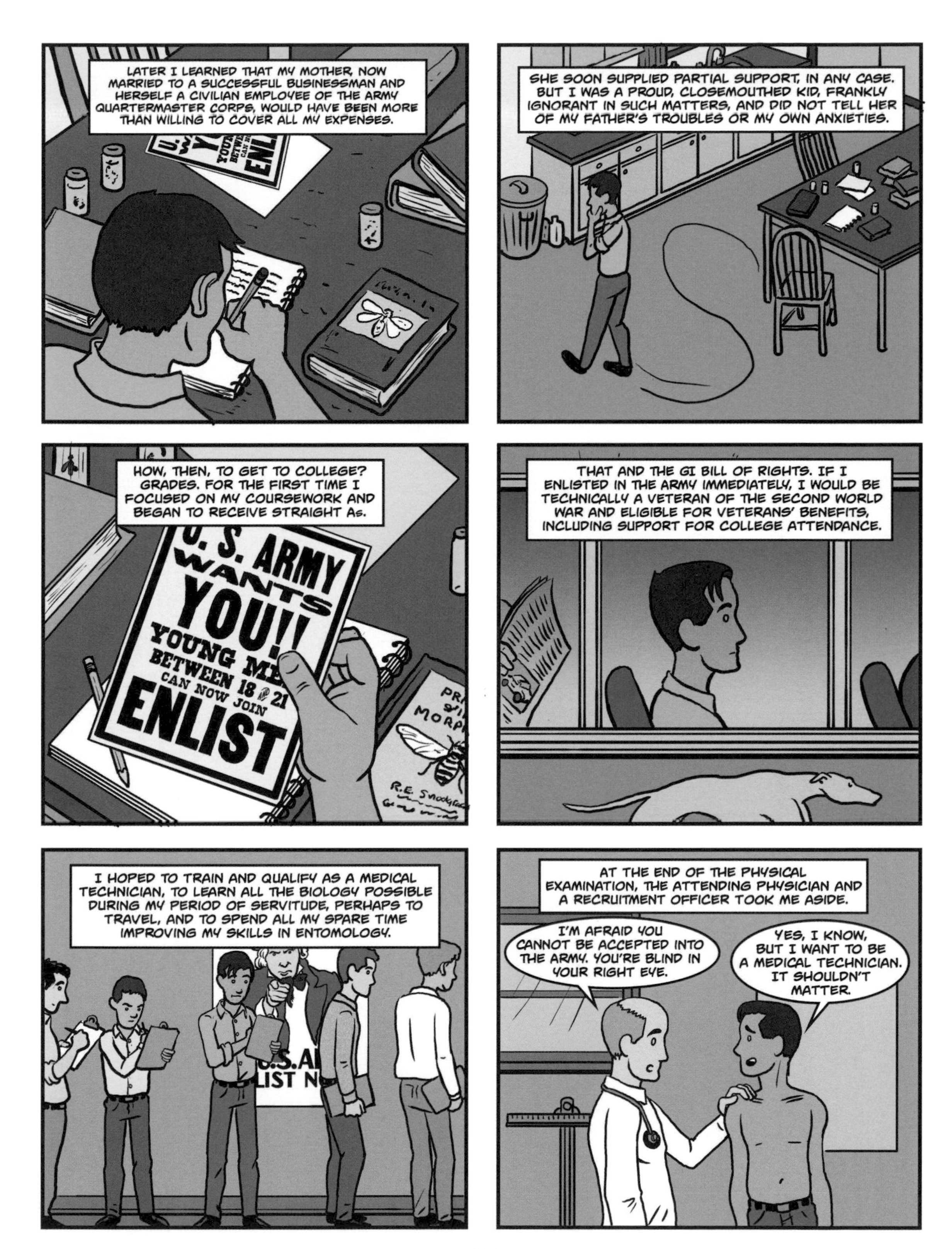
LATER I LEARNED THAT MY MOTHER, NOW MARRIED TO A SUCCESSFUL BUSINESSMAN AND HERSELF A CIVILIAN EMPLOYEE OF THE ARMY QUARTERMASTER CORPS, WOULD HAVE BEEN MORE THAN WILLING TO COVER ALL MY EXPENSES.
SHE SOON SUPPLIED PARTIAL SUPPORT, IN ANY CASE. BUT I WAS A PROUD, CLOSEMOUTHED KID, FRANKLY IGNORANT IN SUCH MATTERS, AND DID NOT TELL HER OF MY FATHER'S TROUBLES OR MY OWN ANXIETIES.
HOW, THEN, TO GET TO COLLEGE? GRADES. FOR THE FIRST TIME I FOCUSED ON MY COURSEWORK AND BEGAN TO RECEIVE STRAIGHT As.
U. S. ARMY WANTS YOU!! YOUNG MEN BETWEEN 18 AND 21 CAN NOW JOIN ENLIST
THAT AND THE GI BILL OF RIGHTS. IF I ENLISTED IN THE ARMY IMMEDIATELY, I WOULD BE TECHNICALLY A VETERAN OF THE SECOND WORLD WAR AND ELIGIBLE FOR VETERANS' BENEFITS, INCLUDING SUPPORT FOR COLLEGE ATTENDANCE.
I HOPED TO TRAIN AND QUALIFY AS A MEDICAL TECHNICIAN, TO LEARN ALL THE BIOLOGY POSSIBLE DURING MY PERIOD OF SERVITUDE, PERHAPS TO TRAVEL, AND TO SPEND ALL MY SPARE TIME IMPROVING MY SKILLS IN ENTOMOLOGY.
AT THE END OF THE PHYSICAL EXAMINATION, THE ATTENDING PHYSICIAN AND A RECRUITMENT OFFICER TOOK ME ASIDE.
I'M AFRAID YOU CANNOT BE ACCEPTED INTO THE ARMY. YOU'RE BLIND IN YOUR RIGHT EYE.
YES, I KNOW, BUT I WANT TO BE A MEDICAL TECHNICIAN. IT SHOULDN'T MATTER.

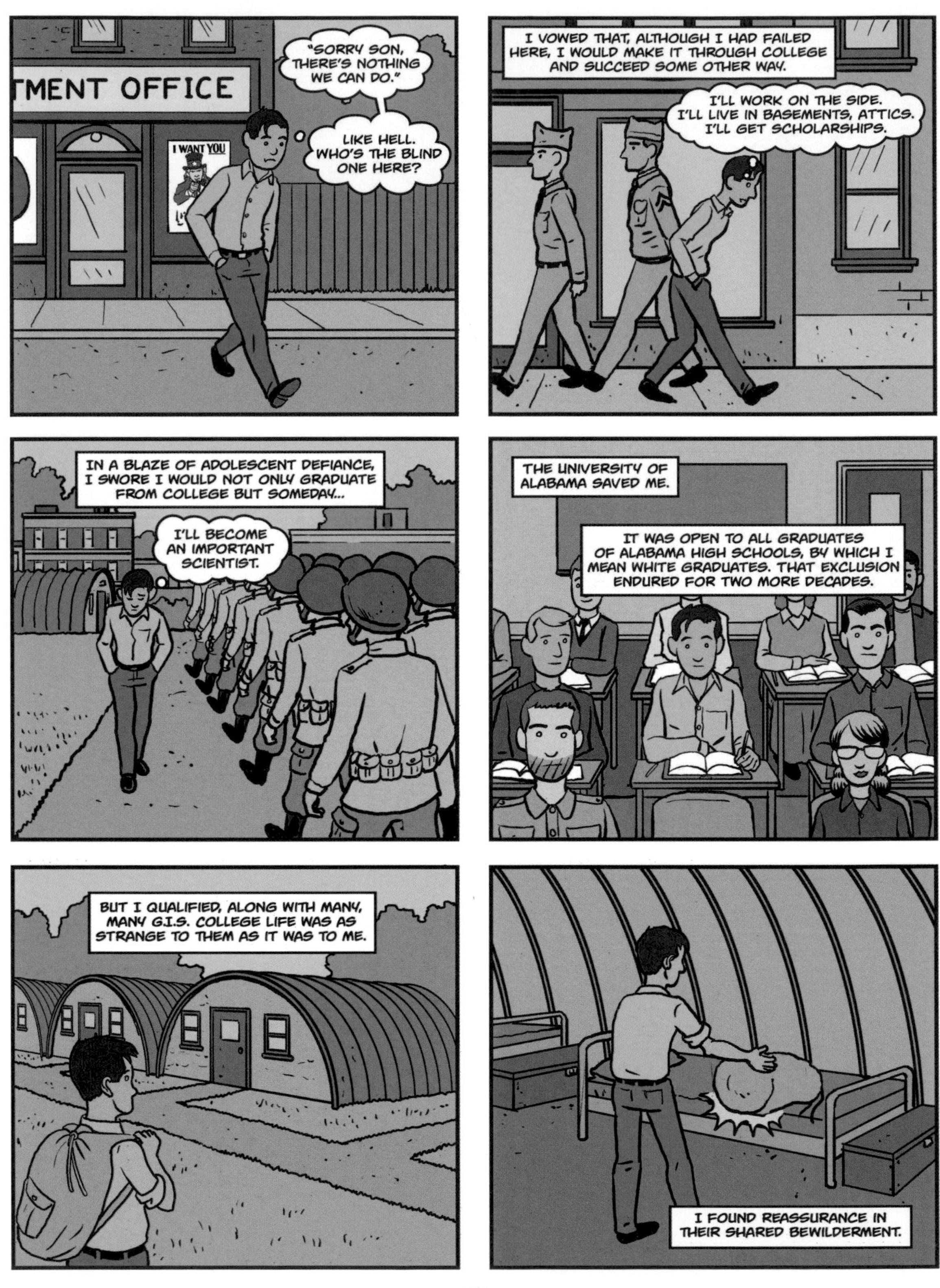
TMENT OFFICE
I WANT YOU
"SORRY SON, THERE'S NOTHING WE CAN DO."
LIKE HELL. WHO'S THE BLIND ONE HERE?
I VOWED THAT, ALTHOUGH I HAD FAILED HERE, I WOULD MAKE IT THROUGH COLLEGE AND SUCCEED SOME OTHER WAY.
I'LL WORK ON THE SIDE. I'LL LIVE IN BASEMENTS, ATTICS. I'LL GET SCHOLARSHIPS.
IN A BLAZE OF ADOLESCENT DEFIANCE, I SWORE I WOULD NOT ONLY GRADUATE FROM COLLEGE BUT SOMEDAY...
I'LL BECOME AN IMPORTANT SCIENTIST.
THE UNIVERSITY OF ALABAMA SAVED ME.
IT WAS OPEN TO ALL GRADUATES OF ALABAMA HIGH SCHOOLS, BY WHICH I MEAN WHITE GRADUATES. THAT EXCLUSION ENDURED FOR TWO MORE DECADES.
BUT I QUALIFIED, ALONG WITH MANY, MANY G.I.S. COLLEGE LIFE WAS AS STRANGE TO THEM AS IT WAS TO ME.
I FOUND REASSURANCE IN THEIR SHARED BEWILDERMENT.

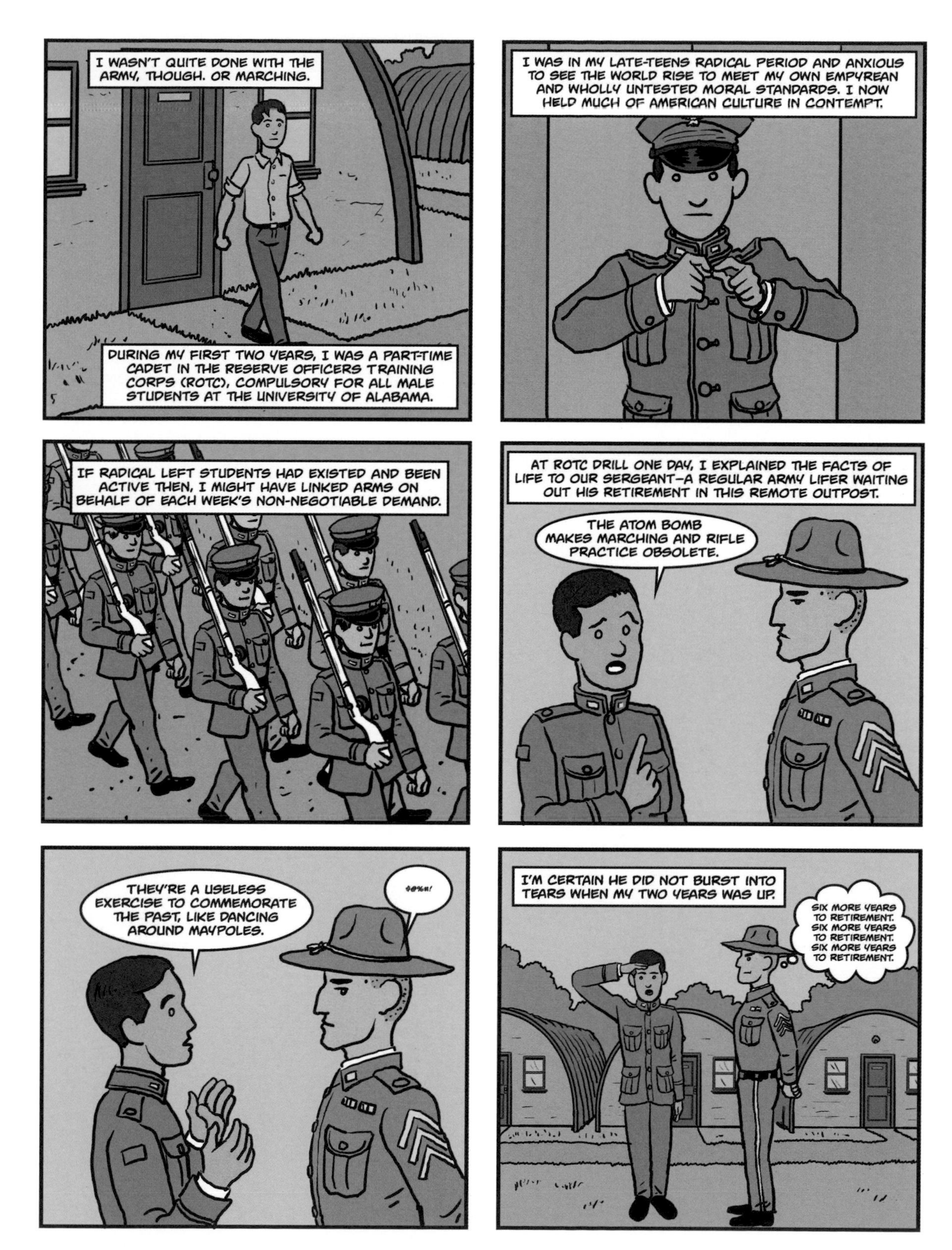
I WASN'T QUITE DONE WITH THE ARMY, THOUGH. OR MARCHING.
DURING MY FIRST TWO YEARS, I WAS A PART-TIME CADET IN THE RESERVE OFFICERS TRAINING CORPS (ROTC), COMPULSORY FOR ALL MALE STUDENTS AT THE UNIVERSITY OF ALABAMA.
I WAS IN MY LATE-TEENS RADICAL PERIOD AND ANXIOUS TO SEE THE WORLD RISE TO MEET MY OWN EMPYREAN AND WHOLLY UNTESTED MORAL STANDARDS. I NOW HELD MUCH OF AMERICAN CULTURE IN CONTEMPT.
IF RADICAL LEFT STUDENTS HAD EXISTED AND BEEN ACTIVE THEN, I MIGHT HAVE LINKED ARMS ON BEHALF OF EACH WEEK'S NON-NEGOTIABLE DEMAND.
AT ROTC DRILL ONE DAY, I EXPLAINED THE FACTS OF LIFE TO OUR SERGEANT—A REGULAR ARMY LIFER WAITING OUT HIS RETIREMENT IN THIS REMOTE OUTPOST.
THE ATOM BOMB MAKES MARCHING AND RIFLE PRACTICE OBSOLETE.
THEY'RE A USELESS EXERCISE TO COMMEMORATE THE PAST, LIKE DANCING AROUND MAYPOLES.
$@%#!
I'M CERTAIN HE DID NOT BURST INTO TEARS WHEN MY TWO YEARS WAS UP.
SIX MORE YEARS TO RETIREMENT. SIX MORE YEARS TO RETIREMENT. SIX MORE YEARS TO RETIREMENT.

SHORTLY AFTER CLASSES STARTED, I CALLED ON PROFESSOR J. HENRY WALKER, HEAD OF THE DEPARTMENT OF BIOLOGY, TO INTRODUCE MYSELF AND TO DISCUSS MY CAREER PLANS.
NOTT HALL
DAY ELLIS AND I CLAMBE ER A STEEP WOODED SLOPE IN PARK, AND I PULLED AWAY THE BARK OF A ROTTING TREE STUMP AND DIS-COVERED A SEETHING MASS OF CIT RONELLA ANTS UNDERNEATH, ME OF THE GENUS ACANTHO
VERY INTERESTING, YOU'VE DONE VERY WELL.
ND A MARCHING COLUMN O TS IN MY BACKYARD—NOT THE FAMO VORACIOUS HORDES OF SOUTH AMERICA RAIN FORESTS, BUT MINIATURE ARMY ANTS OF THE GENUS NEIVAMYRMEX , WHOSE COL ONIES OF 10,000 TO 100,000 WORKERS SEARCH FOR PREY THROUGH GRASS AROUND HUMAN HABITATIO
YES, YES, VERY INTERESTING, FELLA.
ALL YOUNG MEN, IT TURNED OUT, WERE CALLED "FELLA."
NEXT TO THE SIDEWALK ONY WAS SWARMING. THE ANTS POURE F A MOUND NEST, WINGED QUEENS AND MAL TOOK OFF ON THEIR NUPTIAL FLIGHT, PROTECTE BY ANGRY-LOOKING WORKERS THAT RUN UP AND DOWN THE GRASS BLADES AND OUT ONTO THE LISTERING-HOT CONCRETE OF THE SIDEWAL SPECIES WAS UNMISTAKABLY SOLENO MINATA, THE NATIVE FIRE
HE MURMURED REINFORCEMENT AS THOUGH IT WERE ENTIRELY ROUTINE FOR FRESHMAN STUDENTS TO LAUNCH ENTOMOLOGICAL CAREERS IN HIS OFFICE.
BERT, THIS IS ED WILSON. I DO BELIEVE HE'S INTERESTED IN ANTS.
ED, THIS IS PROFESSOR BERT WILLIAMS.
HI ED.
UM. YES. HELLO.
I'M INTERESTED IN ANTS.
SO I GATHERED.

AFTER WE TALKED ANTS, NATURAL HISTORY, AND BOTANY FOR A WHILE, HE TOOK ME TO A TABLE SPACE IN HIS LABORATORY WHERE, HE SUGGESTED, I MIGHT WISH TO CONDUCT MY RESEARCH.
RESEARCH LAB

...
WOW.
HIS LARGESS KNEW NO BOUNDS THEREAFTER.

HE OFFERED TO TAKE ME ALONG ON FUTURE FIELD TRIPS. LATER IN THE YEAR, HE GAVE ME A PART-TIME RESEARCH ASSISTANTSHIP, TRACING RADIOACTIVE PHOSPHORUS THROUGH THE ROOTS OF PLANTS.

I FLOURISHED UNDER THE GUIDANCE OF THESE MULTIPLE ELDERS. IN ADDITION TO TRAINING, THEY GAVE ME THE MOST PRICELESS GIFTS AN APPRENTICE CAN RECEIVE...

DO YOU SEE WHAT I MEAN?
HEY, YES! I DON'T KNOW WHAT THAT'S ABOUT EITHER. HOW ABOUT YOU DO SOME MORE DISSECTIONS, AND LET ME KNOW WHAT YOU THINK?

THEY LET ME KNOW THAT THEY DIDN'T UNDERSTAND EVERYTHING, THAT I MIGHT ACQUIRE INFORMATION THEY DIDN'T HAVE, AND THAT MY EFFORTS WERE VALUED.
I'LL HIT THE LIBRARY. I MAY NOT BE CAUGHT UP ON THE LATEST RESEARCH, SO I'LL LET YOU KNOW WHAT I LEARN.

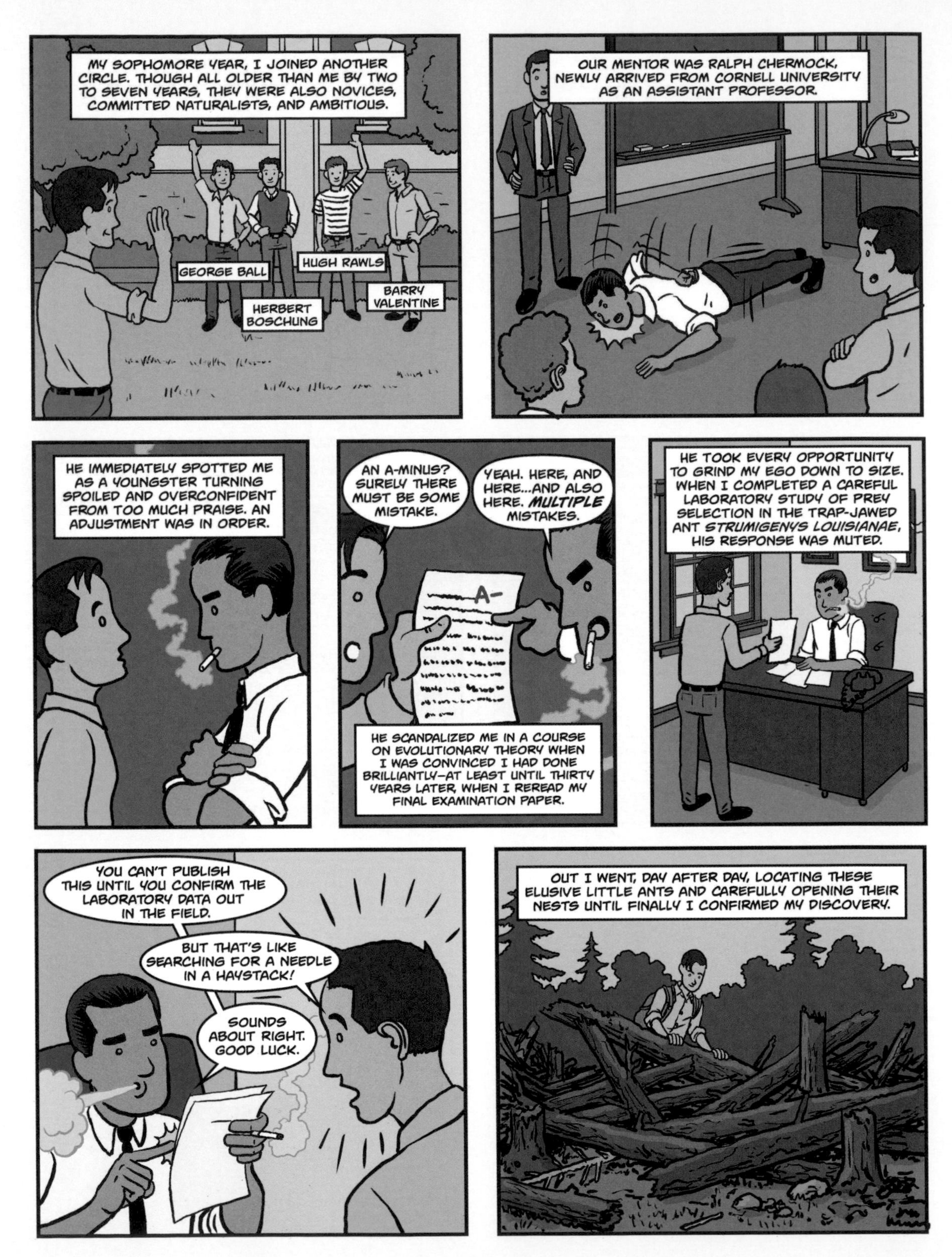

MY SOPHOMORE YEAR, I JOINED ANOTHER CIRCLE. THOUGH ALL OLDER THAN ME BY TWO TO SEVEN YEARS, THEY WERE ALSO NOVICES, COMMITTED NATURALISTS, AND AMBITIOUS.
GEORGE BALL
HERBERT BOSCHUNG
HUGH RAWLS
BARRY VALENTINE
OUR MENTOR WAS RALPH CHERMOCK, NEWLY ARRIVED FROM CORNELL UNIVERSITY AS AN ASSISTANT PROFESSOR.
HE IMMEDIATELY SPOTTED ME AS A YOUNGSTER TURNING SPOILED AND OVERCONFIDENT FROM TOO MUCH PRAISE. AN ADJUSTMENT WAS IN ORDER.
AN A-MINUS? SURELY THERE MUST BE SOME MISTAKE.
YEAH. HERE, AND HERE...AND ALSO HERE. MULTIPLE MISTAKES.
A-
HE SCANDALIZED ME IN A COURSE ON EVOLUTIONARY THEORY WHEN I WAS CONVINCED I HAD DONE BRILLIANTLY—AT LEAST UNTIL THIRTY YEARS LATER, WHEN I REREAD MY FINAL EXAMINATION PAPER.
HE TOOK EVERY OPPORTUNITY TO GRIND MY EGO DOWN TO SIZE. WHEN I COMPLETED A CAREFUL LABORATORY STUDY OF PREY SELECTION IN THE TRAP-JAWED ANT STRUMIGENYS LOUISIANAE, HIS RESPONSE WAS MUTED.
YOU CAN'T PUBLISH THIS UNTIL YOU CONFIRM THE LABORATORY DATA OUT IN THE FIELD.
BUT THAT'S LIKE SEARCHING FOR A NEEDLE IN A HAYSTACK!
SOUNDS ABOUT RIGHT. GOOD LUCK.
OUT I WENT, DAY AFTER DAY, LOCATING THESE ELUSIVE LITTLE ANTS AND CAREFULLY OPENING THEIR NESTS UNTIL FINALLY I CONFIRMED MY DISCOVERY.

AND CHERMOCK RELENTED.
OKAY. CONFIRMED, I SUPPOSE.

THE SEVERAL BEST TEACHERS OF MY LIFE HAVE BEEN THOSE WHO TOLD ME THAT MY VERY BEST WAS NOT YET GOOD ENOUGH.

CHERMOCK WAS A SPECIALIST IN BUTTERFLY CLASSIFICATION AND COMMITTED TO RESEARCH ON EVOLUTIONARY BIOLOGY AND THE MODERN SYNTHESIS OF EVOLUTIONARY THEORY–ESSENTIALLY THE DARWINIAN THEORY OF NATURAL SELECTION WITH MUTATING GENES ADDED.

OUR GROUP WAS THUS EQUIPPED WITH THE TEXTS OF RADICAL AUTHORITY...

...FIELD GUIDES AND OUR OWN PREVIOUSLY ACQUIRED EXPERTISE.

PROVIDENCE SHONE BRIGHT ON ALL OF US TOGETHER—VALENTINE HAD AN AUTOMOBILE.
WE WERE SCIENTIFICALLY LICENSED HUNTERS, WITH THE MEANS TO ROAM AN ECOLOGICALLY DIVERSE STATE THAT HAD UNTIL THEN BEEN ONLY PARTLY EXPLORED BY NATURALISTS.
WE TALKED OF MANY THINGS...
DRYINIDS, PERLIDS, LIMULODIDS...
LIMULODIDS, ENTOMOBRYOMORPHS, PLETHODONTIDS...
PLETHODONTIDS, LITHOBIIDS, SPHINGIDS...
CHERMOCK ENCOURAGED US TO COLLECT NOT ONLY OUR FAVORED ORGANISMS BUT ALSO AMPHIBIANS AND REPTILES FOR THE UNIVERSITY OF ALABAMA COLLECTION.
TZZZZZTZZZZZZZZTZZZZZZTZZZZZ

SPHINGIDS, LIBELLUDIDS
WAIT!
TZZZZZZZZZZZZZTZZZZZTZZZZZ
HEY, DO YOU HEAR THAT OVER ON THE LEFT?
NO, BUT I THINK I SEE...
ONE NIGHT WE DROVE SLOWLY INTO THE FLORIDA PANHANDLE. WE WERE SEARCHING FOR THE ZONE WHERE THE NORTHERN RACE *PSEUDACRIS NIGRITA TRISERIATA* MEETS AND INTERBREEDS WITH THE SOUTHERN RACE, *PSEUDACRIS NIGRITA NIGRITA*.
TZZZZZTZZZZZZZZTZZZZZZZTZZZZZ
THEY SING WITH DIFFERENT TRILL PATTERNS. FOR A CLOSE APPROXIMATION OF A CHORUS FROG CALL, RUN THE EDGE OF YOUR FINGERNAIL ALONG THE FINE TEETH OF A POCKET COMB.
STILL CAN'T HEAR IT. SORRY.
TZZZZZTZZZZZZZZTZZZZZZZTZZZZZ
NEAR DAWN, WE ENCOUNTERED THE CHANGEOVER CLOSE TO THE FLORIDA BORDER, AND THEN IT PROVED TO BE VERY ABRUPT.
WE REASONED THAT THE TWO FORMS ARE ACTUALLY REPRODUCTIVELY ISOLATED SPECIES, NOT INTERBREEDING RACES, AND DESERVED THEIR FORMAL DISTINCTION AS *PSEUDACRIS TRISERIATA* AND *PSEUDACRIS NIGRITA*. RESEARCH BY LATER SPECIALISTS PROVED US RIGHT.

YOU DON'T HAVE TO GO FAR TO DISCOVER THINGS. ON OTHER NIGHTS, WE SIMPLY WALKED THE STREETS OF TUSCALOOSA. SCIENTIFIC DISCOVERY AT THIS ELEMENTARY LEVEL WAS ALL SO EASY, ALL SUCH FUN.

DURING THESE EXPEDITIONS, I SOAKED UP NEW INFORMATION—ON AND ON DEEP INTO THE HEART OF BIODIVERSITY.

I COULD NOT UNDERSTAND WHY MOST OF THE OTHER STUDENTS AT THE UNIVERSITY DID NOT ALSO ASPIRE TO BE BIOLOGISTS.

CHERMOCK WAS UNIMPRESSED BY OUR GROWING EXPERTISE.
YOU CAN'T CALL YOURSEVES BIOLOGISTS UNTIL YOU KNOW THE NAMES OF 10,000 KINDS OF ORGANISMS.

DO YOU THINK HE CAN PASS THAT TEST?
NAH. BUT IT DOESN'T MATTER.
IT DIDN'T. HYPERBOLE FROM THE CHIEF KEPT OUR JUICES FLOWING.

SO, BY THE AGE OF EIGHTEEN, I HAD BEEN CONVERTED TO SCIENTIFIC PROFESSIONALISM. BARELY OUT OF MY BOY SCOUT YEARS, I WAS BACK ON THE TRAIL OF MERIT BADGES, THIS TIME THROUGH RESEARCH, DISCOVERY, PUBLICATION...

AND, AS ALFRED NORTH WHITEHEAD ONCE SAID OF SCIENTISTS GENERALLY, I DID NOT DISCOVER IN ORDER TO LEARN...I LEARNED IN ORDER TO DISCOVER.
MEANWHILE, I DEVELOPED A STRONG NEW RESEARCH INTEREST IN THE IMPORTED FIRE ANT, WHICH I HAD FIRST OBSERVED IN MOBILE IN 1942.
SOLENOPSIS SAEVISSIMA VAR. RICHTERI
AS A RESULT, IN EARLY 1949, THE ALABAMA DEPARTMENT OF CONSERVATION ASKED ME TO CONDUCT A STUDY OF THE ANT.
...NEED YOU TO EVALUATE ITS IMPACT ON THE ENVIRONMENT.
I TOOK LEAVE FROM THE UNIVERSITY FOR THE SPRING TERM TO BEGIN, AT THE AGE OF NINETEEN, A FOUR-MONTH STINT AS ENTOMOLOGIST, MY FIRST POSITION AS A PROFESSIONAL SCIENTIST.
I WAS JOINED BY JAMES EADS, ANOTHER BIOLOGIST.
LIKE MY OTHER COMPANIONS, HE WAS A WAR VETERAN IN HIS MID-TWENTIES AND, MOST CRUCIALLY AGAIN, OWNER OF A CAR.
WE CRISSCROSSED THE TERRITORY, MAPPING THE EXPANDING SEMICIRCULAR RANGE OF THE ANT.

IT LOOKS LIKE THEY PUSH UP SOIL AS THEY TUNNEL DOWN.

WHEN DID THEY FIRST APPEAR HERE?

IN JULY, WE SUBMITTED A FIFTY-THREE-PAGE ANALYSIS TO THE DEPARTMENT OF CONSERVATION TITLED "A REPORT ON THE IMPORTED FIRE ANT *SOLENOPSIS SAEVISSIMA* VAR. *RICHTERI FOREL* IN ALABAMA."

IT CONTAINED ORIGINAL FINDINGS ON THE ANT STILL IN USE TODAY, INCLUDING THE RATE OF SPREAD (FIVE MILES A YEAR ALONG ALL BORDERS), THE PARTIAL ELIMINATION OF NATIVE FIRE ANT SPECIES, AND DOCUMENTATION OF MODERATE CROP DAMAGE.

EVEN WITH THIS DETOUR, I FINISHED MY BACHELOR OF SCIENCE DEGREE IN THREE YEARS, SPENDING ONLY A LITTLE OVER $2,000, SOMEWHAT LESS THAN THE ANNUAL SALARY OF A GOVERNMENT CLERK OR SCHOOLTEACHER AT THE TIME.

IN 1950, I TRANSFERRED TO THE UNIVERSITY OF TENNESSEE TO BEGIN WORK ON MY PH.D., MAINLY BECAUSE OF THE PRESENCE THERE OF ARTHUR COLE, A PROFESSOR OF ENTOMOLOGY WHO SPECIALIZED IN THE CLASSIFICATION OF ANTS.

BUT THE ACADEMIC CHALLENGE WAS NOT GREAT, AND OUT OF BOREDOM, I BECAME A BIT RECKLESS.

IN THE FALL OF THAT YEAR, I LEARNED ABOUT THE EXTRAORDINARY DISCOVERY OF THE FIRST OF THE ERECT, SMALL-BRAINED HOMINIDS.

THE MISSING LINKS BETWEEN REMOTE APE-LIKE ANCESTORS AND THE MOST PRIMITIVE TRUE HUMANS OF THE GENUS HOMO. HERE, I THOUGHT, WAS ONE OF THE MOST IMPORTANT SCIENTIFIC DISCOVERIES OF THE CENTURY.

I WAS INTRIGUED THAT THE TEACHING OF EVOLUTION IN THE STATE WAS FORBIDDEN.
NOT SURE IT'S ENFORCED, BUT... THERE *IS* A STATUTE MAKING IT AGAINST THE LAW, ED.

I WAS ALSO INTRIGUED BY ITS DEEP SIGNIFICANCE FOR THE SELF-IMAGE OF OUR SPECIES.

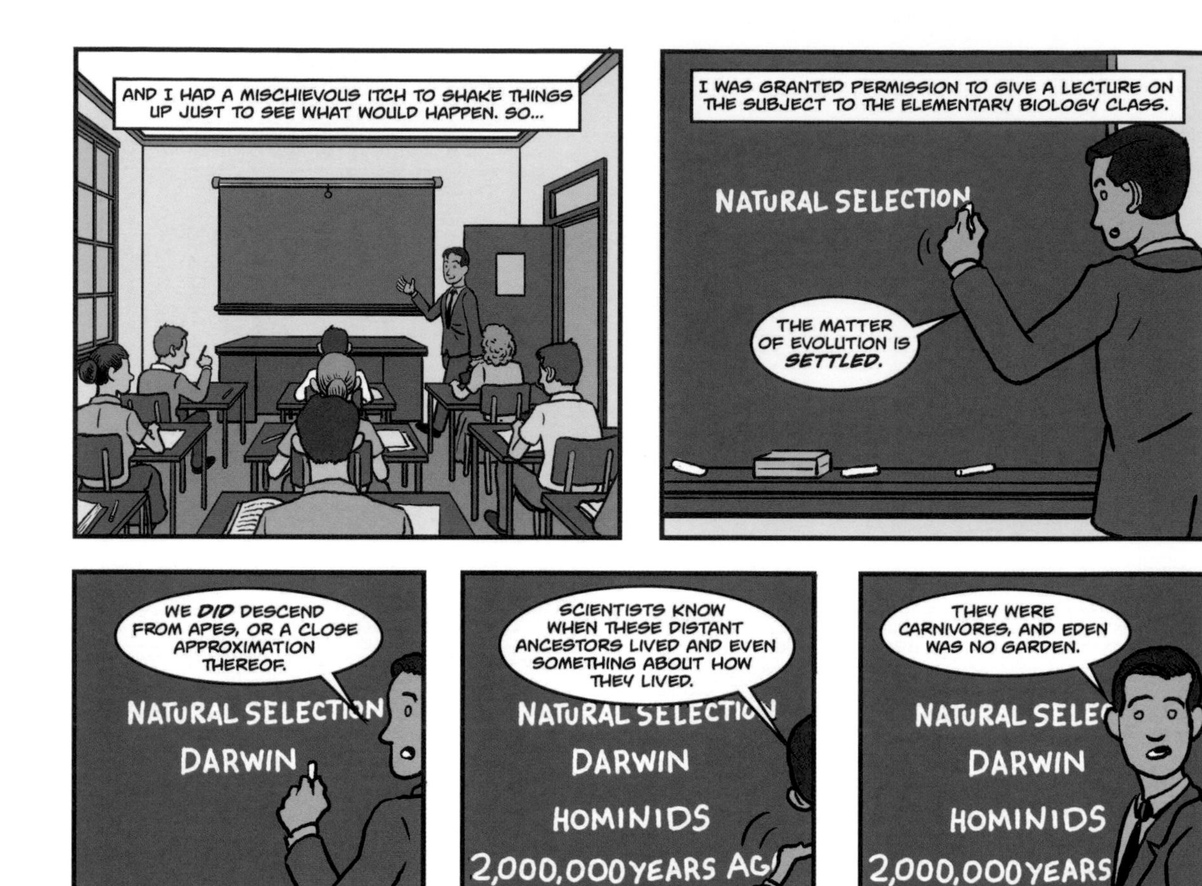

THE STUDENTS WERE MOSTLY PROTESTANTS, AND MANY HAD BEEN RAISED IN FUNDAMENTALIST FAMILIES. SOME, I'M SURE, HAD BEEN TAUGHT THAT DARWIN WAS THE DEVIL'S PARSON.

THIS EVOLUTION STUFF. I NEED YOU TO TELL ME JUST ONE THING.
...
WILL THIS BE ON THE FINAL EXAM?
...
UM, NO. PLEASE DON'T WORRY.
WHEW. THANKS. ONE LESS THING TO MEMORIZE!
I LEARNED A LESSON OF MY OWN—THE GREATER PROBLEMS OF HISTORY ARE NOT SOLVED.
THEY ARE MERELY FORGOTTEN.

BY EARLY 1951, I HAD DECIDED TO MOVE ON TO HARVARD UNIVERSITY.
IT WAS MY DESTINY.

THEY HAVE THE LARGEST COLLECTION OF ANTS IN THE WORLD, AND A LONG AND DEEP TRADITION OF STUDYING THEM.

I KNOW, BUT I DON'T THINK I CAN AFFORD...
I'M NOMINATING YOU FOR A FELLOWSHIP.
A SECOND SUPPORTER WAS WILLIAM L. BROWN, THEN A GRADUATE STUDENT IN HARVARD'S DEPARTMENT OF BIOLOGY.

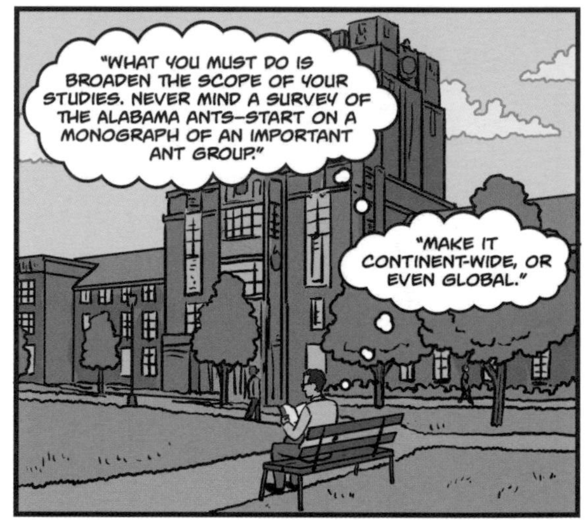
"WHAT YOU MUST DO IS BROADEN THE SCOPE OF YOUR STUDIES. NEVER MIND A SURVEY OF THE ALABAMA ANTS—START ON A MONOGRAPH OF AN IMPORTANT ANT GROUP."
"MAKE IT CONTINENT-WIDE, OR EVEN GLOBAL."

"RIGHT NOW, YOU HAVE THE ADVANTAGE OF LIVING IN THE DEEP SOUTH, WHERE THERE ARE A GREAT MANY DACETINE ANTS. THESE ARE EXTRAORDINARILY INTERESTING INSECTS, AND WE STILL DON'T KNOW MUCH ABOUT THEM."

HE CLOSED WITH "THERE IS AN OPPORTUNITY TO DO SOME REALLY ORIGINAL RESEARCH. SEE WHAT YOU CAN COME UP WITH..."

I PLUNGED INTO THE DACETINE PROJECT AT ONCE, TRACKING DOWN SPECIES, ONE AFTER ANOTHER.
LARGELY BECAUSE I ENJOYED GRUBBING IN DIRT AND ROTTING WOOD ON HANDS AND KNEES, I WAS VERY SUCCESSFUL IN MY PURSUIT.
THE DACETINES ARE SLENDER, ORNATELY SCULPTURED LITTLE ANTS WITH LONG, THIN MANDIBLES.
CLEAN AND DECORATIVE, THEY ARE AMONG THE MOST AESTHETICALLY PLEASING OF ALL INSECTS.

BROWN, OR UNCLE BILL AS HE WAS TO BECOME AFFECTIONATELY KNOWN BY YOUNGER ENTOMOLOGISTS, URGED ME TO VISIT HARVARD. I DID SO IN LATE JUNE 1950.

ED? BILL BROWN.
YOU LOOK... REALLY TIRED.
I THINK WE STOPPED IN EVERY MID-SIZED TOWN BETWEEN MOBILE AND BOSTON. SO, YES, I AM.

DURING THE NEXT SEVERAL DAYS, EVEN AS BILL MADE FINAL PREPARATIONS TO LEAVE FOR FIELD WORK IN AUSTRALIA, HE TOOK TIME TO GUIDE ME THROUGH HARVARD...

... AND ITS ANT COLLECTION.
YOUR DACETINE AND FIRE ANT STUDIES ARE VERY PROMISING.
BUT, NOW YOU SHOULD COME TO HARVARD. BIGGER PROJECTS, GREATER SCOPE.

TAKE A GLOBAL VIEW.
I DID.

AND LATER, WE DID. HE AND I CORRELATED THE FOOD HABITS OF LARGE NUMBERS OF SPECIES FROM AROUND THE WORLD WITH THEIR SOCIAL ORGANIZATION.
THE MORE ANATOMICALLY PRIMITIVE THE ANT, THE MORE CASTES IN A COLONY...

OURS WAS A NOVEL APPROACH TO THE STUDY OF BEHAVIOR. SO FAR AS I AM AWARE, THE STUDY WAS THE FIRST OF ITS KIND ON THE EVOLUTION OF SOCIAL ECOLOGY IN ANIMALS.
...AND THE MORE DIVERSE THEIR FOOD SOURCE.

MORE ADVANCED ANTS HAD SMALLER, MORE UNIFORM BODIES, LESS DIVISION OF LABOR, AND MORE SPECIALIZED FOOD SOURCES. OUR STUDY MIGHT HAVE HAD A HUGE EFFECT ON SOCIOBIOLOGY.
(MORE ON THAT LATER.)
YES. THIS WILL BREAK THINGS OPEN.

IT DIDN'T.
PARTLY BECAUSE MONKEYS, BIRDS, AND OTHER VERTEBRATES ARE MORE NEARLY HUMAN-SIZED AND MORE FAMILIAR, THEY GET TREATED AS MORE "IMPORTANT."
I WAS ADMITTED TO HARVARD FOR THE COMING FALL SEMESTER WITH A SCHOLARSHIP AND TEACHING ASSISTANTSHIP THAT COVERED ALL EXPENSES.
THE SALVATION ARMY
SO, IN AUGUST 1951, I SOLD MY ONLY SUIT FOR TEN DOLLARS, PACKED MY THINGS INTO A SINGLE SUITCASE, AND TRAVELED TO VISIT MY MOTHER AND HER HUSBAND, HAROLD IN LOUISVILLE.
THE SALVATION ARMY
AFTER TAKING ONE LOOK AT ME, HAROLD ESCORTED ME TO A MEN'S CLOTHING STORE AND BOUGHT ME A WARDROBE BEFITTING A 1951 HARVARD STUDENT.
WITH A FRESH CREW CUT ADDED, I WAS READY TO PASS INTO A NEW LIFE. GOOD-BYE TO THE SOUTH.

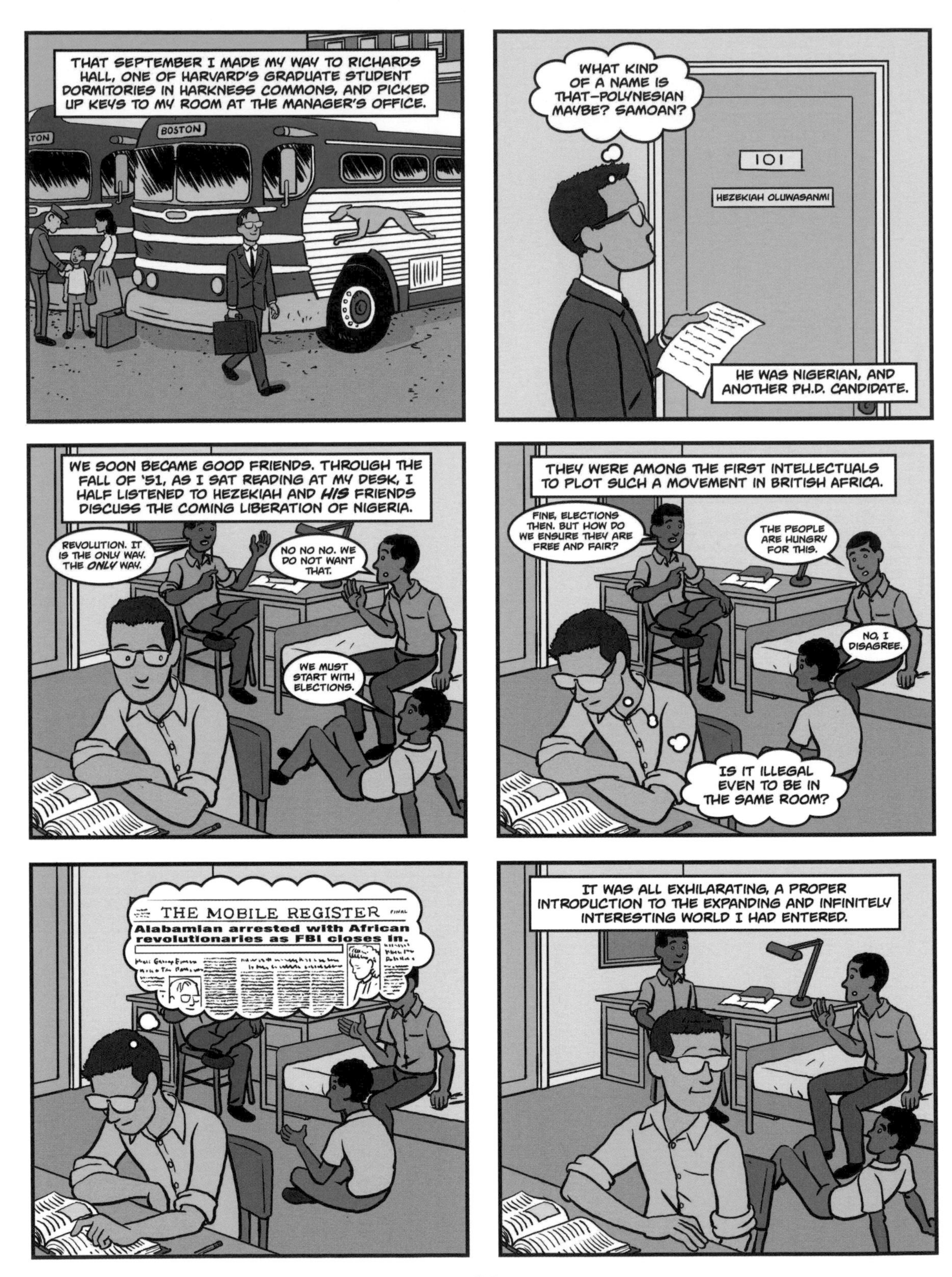
THAT SEPTEMBER I MADE MY WAY TO RICHARDS HALL, ONE OF HARVARD'S GRADUATE STUDENT DORMITORIES IN HARKNESS COMMONS, AND PICKED UP KEYS TO MY ROOM AT THE MANAGER'S OFFICE.
BOSTON
WHAT KIND OF A NAME IS THAT—POLYNESIAN MAYBE? SAMOAN?
101
HEZEKIAH OLUWASANMI
HE WAS NIGERIAN, AND ANOTHER PH.D. CANDIDATE.
WE SOON BECAME GOOD FRIENDS. THROUGH THE FALL OF '51, AS I SAT READING AT MY DESK, I HALF LISTENED TO HEZEKIAH AND *HIS* FRIENDS DISCUSS THE COMING LIBERATION OF NIGERIA.
REVOLUTION. IT IS THE ONLY WAY. THE *ONLY* WAY.
NO NO NO. WE DO NOT WANT THAT.
WE MUST START WITH ELECTIONS.
THEY WERE AMONG THE FIRST INTELLECTUALS TO PLOT SUCH A MOVEMENT IN BRITISH AFRICA.
FINE, ELECTIONS THEN. BUT HOW DO WE ENSURE THEY ARE FREE AND FAIR?
THE PEOPLE ARE HUNGRY FOR THIS.
NO, I DISAGREE.
IS IT ILLEGAL EVEN TO BE IN THE SAME ROOM?
THE MOBILE REGISTER
FINAL
Alabamian arrested with African revolutionaries as FBI closes in.
IT WAS ALL EXHILARATING, A PROPER INTRODUCTION TO THE EXPANDING AND INFINITELY INTERESTING WORLD I HAD ENTERED.

ALMOST ALL MY LIFE I'VE DREAMED OF THE TROPICS.
MY FAVORITE NOVEL WAS ARTHUR CONAN DOYLE'S ***THE LOST WORLD,*** WHICH HINTED THAT LONG-EXTINCT MARVELS MIGHT YET BE FOUND ON THE FLAT SUMMIT OF SOME UNCLIMBED SOUTH AMERICAN TEPUI.
AFTER I ENTERED COLLEGE, I HAD EXPLORED THE EDGES OF THE MOBILE-TENSAW DELTA FLOODPLAIN.
IT WAS A PLACE NO FIELD BIOLOGIST HAD VISITED—AND WAS SELDOM ENTERED BY ANYONE FOR ANY REASON.
MIGHT IT CONTAIN UNDISCOVERED SPECIES LIVING IN ECOLOGICAL NICHES NEW TO SCIENCE? I DECIDED I'D CONDUCT AN EXPEDITION TO INAUGURATE MY CAREER AS A TROPICAL EXPLORER.
IT'S NOT SOUTH AMERICA OR AFRICA, BUT...
I NEVER MADE IT ***INTO*** THE DELTA, THOUGH.
AND IN MY FIRST YEAR AT HARVARD, I WAS DELAYED FURTHER. I SETTLED ON A SENSIBLE THESIS PROJECT THAT COULD BE FINISHED IN THREE OR FOUR YEARS.
TYRANNOMYRMEX REX
ACROPYGA
ATTA CUBANA
CARDIOCONDYLA EMERY
NEIVAMYRMEX NIGRESCENS
NOMAMYRMEX ESENBECKII
PLATYTHYREA PUNCTATA
PROCERATIUM SILAC
STIGMATOMMA P
SYSCIA AUGUSTA
DINOMYRMEX GIGA
LASIUS.
THEN, I FIGURED, I COULD GO TO THE TROPICS.

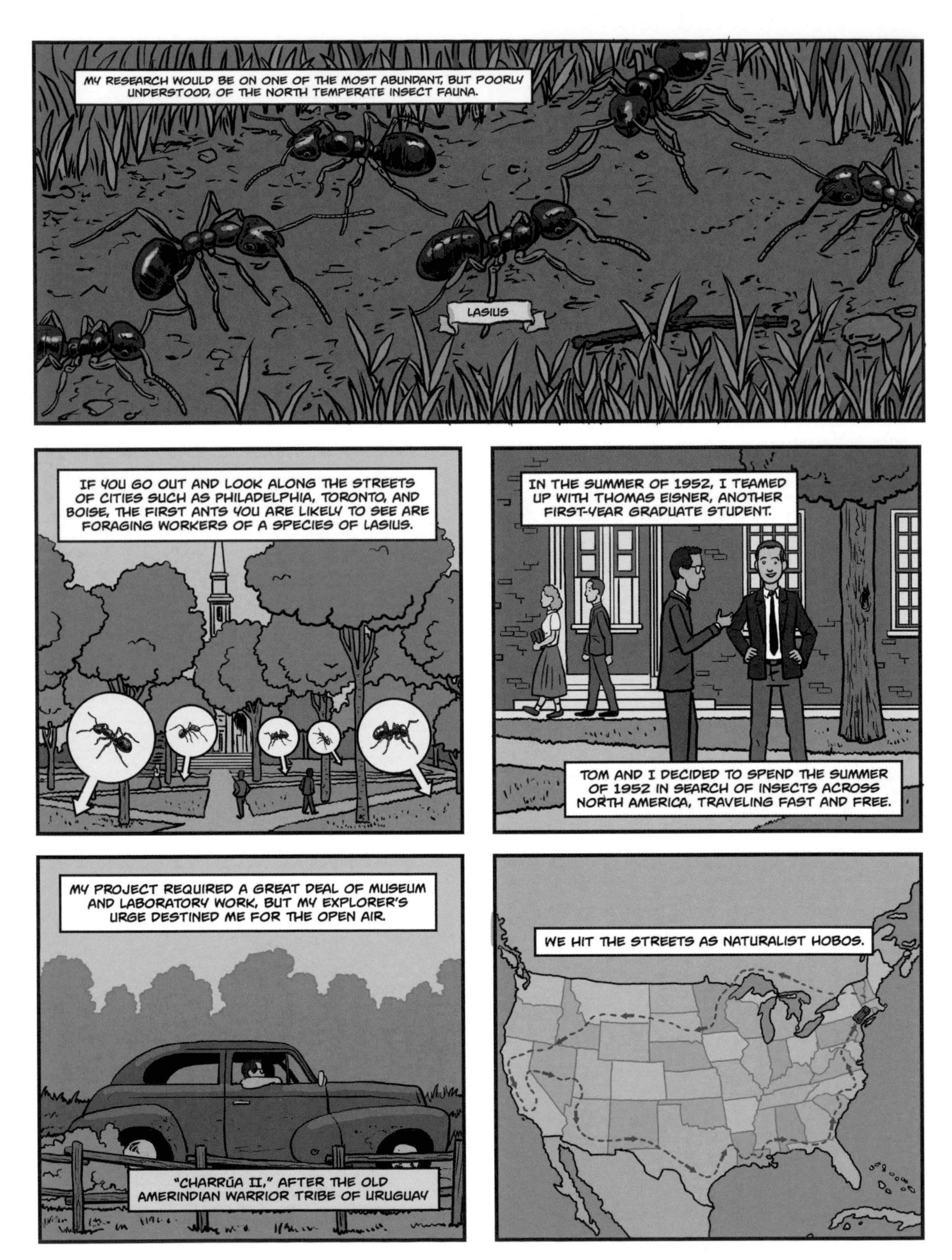
MY RESEARCH WOULD BE ON ONE OF THE MOST ABUNDANT, BUT POORLY UNDERSTOOD, OF THE NORTH TEMPERATE INSECT FAUNA.
LASIUS
IF YOU GO OUT AND LOOK ALONG THE STREETS OF CITIES SUCH AS PHILADELPHIA, TORONTO, AND BOISE, THE FIRST ANTS YOU ARE LIKELY TO SEE ARE FORAGING WORKERS OF A SPECIES OF LASIUS.
IN THE SUMMER OF 1952, I TEAMED UP WITH THOMAS EISNER, ANOTHER FIRST-YEAR GRADUATE STUDENT.
TOM AND I DECIDED TO SPEND THE SUMMER OF 1952 IN SEARCH OF INSECTS ACROSS NORTH AMERICA, TRAVELING FAST AND FREE.
MY PROJECT REQUIRED A GREAT DEAL OF MUSEUM AND LABORATORY WORK, BUT MY EXPLORER'S URGE DESTINED ME FOR THE OPEN AIR.
"CHARRÚA II," AFTER THE OLD AMERINDIAN WARRIOR TRIBE OF URUGUAY
WE HIT THE STREETS AS NATURALIST HOBOS.

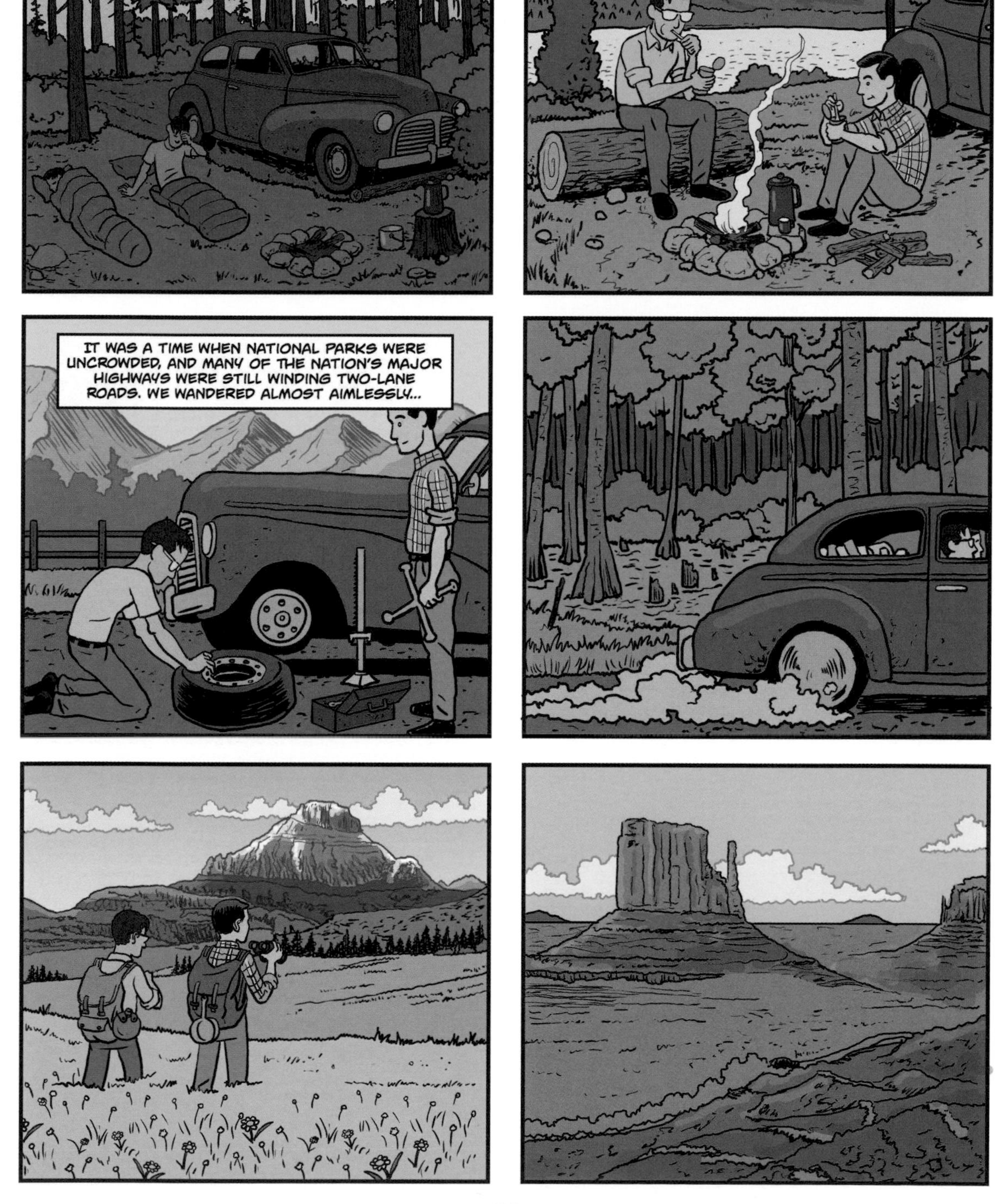
EACH NIGHT, WE SLEPT ON THE GROUND, SOMETIMES IN THE FEE-LESS CAMPING AREAS OF STATE PARKS, MORE OFTEN ON THE EDGE OF OPEN FIELDS AND WOODLOTS OFF THE SIDE OF THE ROAD.
WE PUT MOST OF OUR NEGLIGIBLE FUNDS INTO THE CARE AND FUELING OF CHARRÚA II.
IT WAS A TIME WHEN NATIONAL PARKS WERE UNCROWDED, AND MANY OF THE NATION'S MAJOR HIGHWAYS WERE STILL WINDING TWO-LANE ROADS. WE WANDERED ALMOST AIMLESSLY...

A FEW MONTHS LATER, IN THE SPRING OF 1953, I WAS HANDED THE OPPORTUNITY OF A LIFE-TIME-ELECTION AS A JUNIOR FELLOW IN HARVARD UNIVERSITY'S SOCIETY OF FELLOWS.
I FOUND MYSELF LODGED FREE IN LOWELL HOUSE WITH A GENEROUS STIPEND, A BOOK ALLOWANCE, AND TRAVEL FUNDS AVAILABLE UPON APPLICATION.
IT'S A LONG WAY FROM A PADDED CELL IN THE BARRACKS!
WAIT...WHERE DID YOU GO TO SCHOOL AGAIN?
AT THE FIRST DINNER OF THE FALL TERM, WE NEW FELLOWS STOOD AS THE SOCIETY CHAIRMAN—THE HISTORIAN CRANE BRINTON—READ THE STATEMENT WRITTEN BY ABBOTT LAWRENCE LOWELL.
YOU HAVE BEEN SELECTED AS A MEMBER OF THIS SOCIETY FOR YOUR PERSONAL PROSPECT OF ACHIEVEMENT IN YOUR CHOSEN FIELD, AND YOUR PROMISE OF NOTABLE CONTRIBUTION TO KNOWLEDGE AND THOUGHT.
THAT PROMISE YOU MUST REDEEM WITH YOUR WHOLE INTELLECTUAL AND MORAL FORCE.
YOU WILL SEEK NOT A NEAR, BUT A DISTANT, OBJECTIVE, AND YOU WILL NOT BE SATISFIED WITH WHAT YOU HAVE DONE.
ALL THAT YOU MAY ACHIEVE OR DISCOVER YOU WILL REGARD AS A FRAGMENT OF A LARGER PATTERN, WHICH FROM EACH SEPARATE APPROACH EVERY TRUE SCHOLAR IS STRIVING TO DESCRY.
FAIR ENOUGH. I HAVE THREE YEARS TO JUSTIFY THE CONFIDENCE PLACED IN ME...
THE SAME AMOUNT OF TIME IT TOOK TO MAKE EAGLE SCOUT.

NO PROBLEM
THE SECOND GIFT FROM THE SOCIETY OF FELLOWS WAS TO PLACE ME IN THE WEEKLY COMPANY OF OTHER YOUNG MEN, ALL IN THEIR TWENTIES, WHO HAD BEGUN TO EXCEL IN WIDELY DIVERSE FIELDS OF LEARNING.
HENRY ROSOVSKY (ECONOMIC HISTORIAN)
DONALD HALL (POET)
NOAM CHOMSKY (LINGUIST)
AND IT WASN'T JUST YOUNG MEN. AMONG THE MANY NOTABLE DINNER GUESTS I MET DURING MY THREE YEARS AS A JUNIOR FELLOW WERE T. S. ELIOT, ROBERT OPPENHEIMER, AND ISIDOR I. RABI.
I.I. RABI, NOBEL PRIZE WINNING PHYSICIST
I SPENT AN ESPECIALLY MEMORABLE EVENING ARGUING WITH RABI ABOUT THE EVOLUTIONARY CONSEQUENCES OF ATOMIC BOMB TESTS.
THESE EXPLOSIONS ARE GOOD.
RADIATION INCREASES THE RATE OF MUTATION, WHICH CAN SPEED EVOLUTION.
THAT IS A GOOD THING, IS IT NOT?
!
I DON'T...
YOU CAN'T BE...
IS HE SERIOUS?
I WAS NOT COMPLETELY SURE, BUT THE CONVERSATION WAS INFORMATION-PACKED AND EXCITING EITHER WAY.

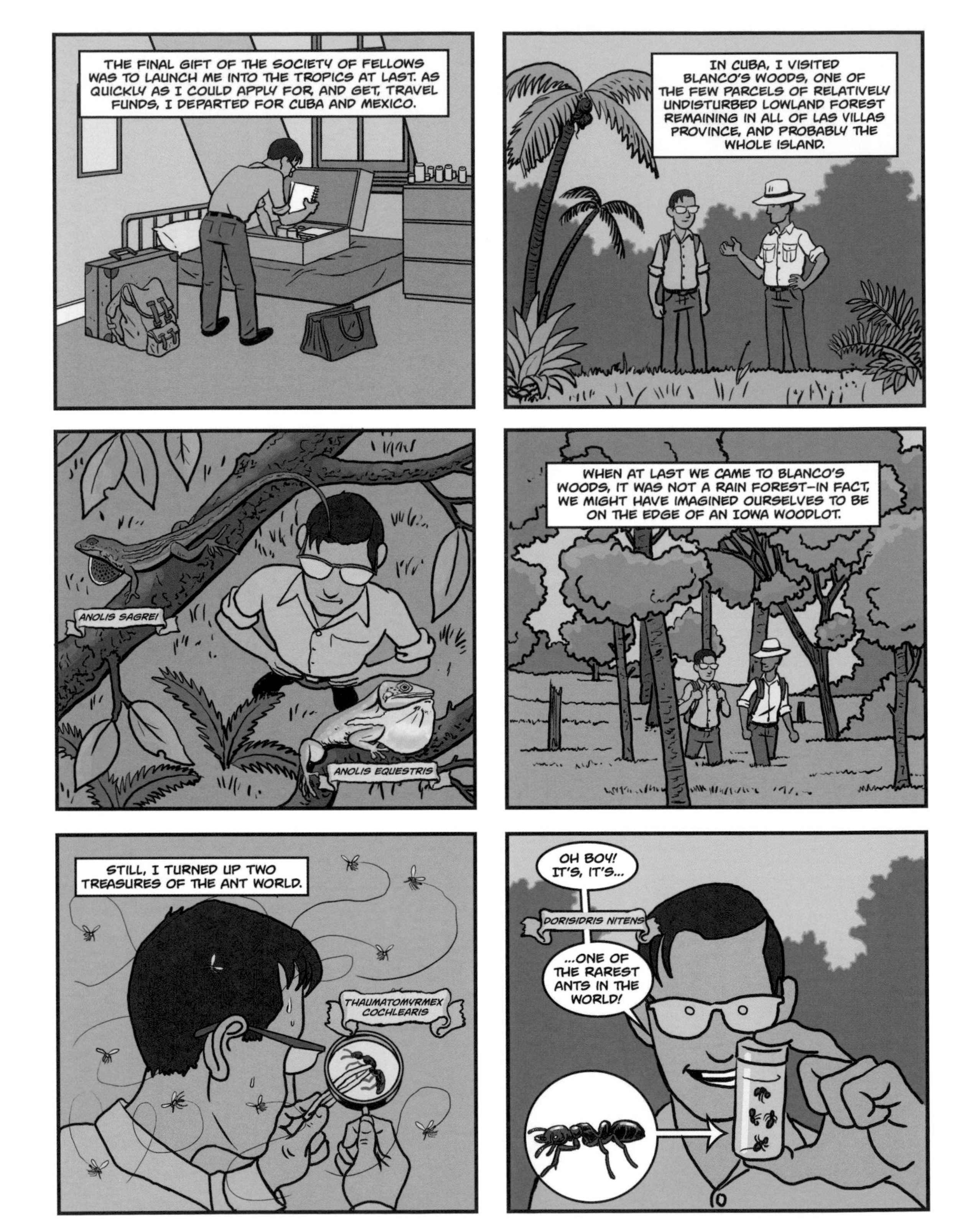
THE FINAL GIFT OF THE SOCIETY OF FELLOWS WAS TO LAUNCH ME INTO THE TROPICS AT LAST. AS QUICKLY AS I COULD APPLY FOR, AND GET, TRAVEL FUNDS, I DEPARTED FOR CUBA AND MEXICO.
IN CUBA, I VISITED BLANCO'S WOODS, ONE OF THE FEW PARCELS OF RELATIVELY UNDISTURBED LOWLAND FOREST REMAINING IN ALL OF LAS VILLAS PROVINCE, AND PROBABLY THE WHOLE ISLAND.
ANOLIS SAGREI
ANOLIS EQUESTRIS
WHEN AT LAST WE CAME TO BLANCO'S WOODS, IT WAS NOT A RAIN FOREST—IN FACT, WE MIGHT HAVE IMAGINED OURSELVES TO BE ON THE EDGE OF AN IOWA WOODLOT.
STILL, I TURNED UP TWO TREASURES OF THE ANT WORLD.
THAUMATOMYRMEX COCHLEARIS
OH BOY! IT'S, IT'S...
DORISIDRIS NITENS
...ONE OF THE RAREST ANTS IN THE WORLD!

¿ESTÁ EN LA COLECCIÓN DE HARVARD?
NOT NOW, BUT IT WILL BE SOON!
UM, AHORA NO, PERO... PRONTO.

ONWARD TO MINA CARLOTA, WHERE WILLIAM MANN—ANOTHER HARVARD GRADUATE STUDENT, THEN DIRECTOR OF THE NATIONAL ZOO—HAD DISCOVERED *MACROMISCHA WHEELERI* FORTY YEARS EARLIER.

AND THAT TOOK ME BACK...
AN UNKNOWN SPECIES OF ANTS LIKE... LIKE...
"...LIKE LIVING EMERALDS." IT'S IN YOUR NATIONAL GEOGRAPHIC ARTICLE!

I TOOK A SPECIAL SATISFACTION IN REPEATING MANN'S DISCOVERY IN EXACT DETAIL AFTER SUCH A LONG INTERVAL.
IT WAS A REASSURANCE OF THE CONTINUITY OF BOTH THE NATURAL WORLD AND THE HUMAN MIND.

ON THE TRINIDAD MASSIF, I ENCOUNTERED ANOTHER ANT WHOSE WORKERS GLISTENED GOLDEN IN THE SUNLIGHT. BRIGHT COLORS ARE WIDESPREAD AMONG THESE MACROMISCHA.
MACROMISCHA SQUAMIFER

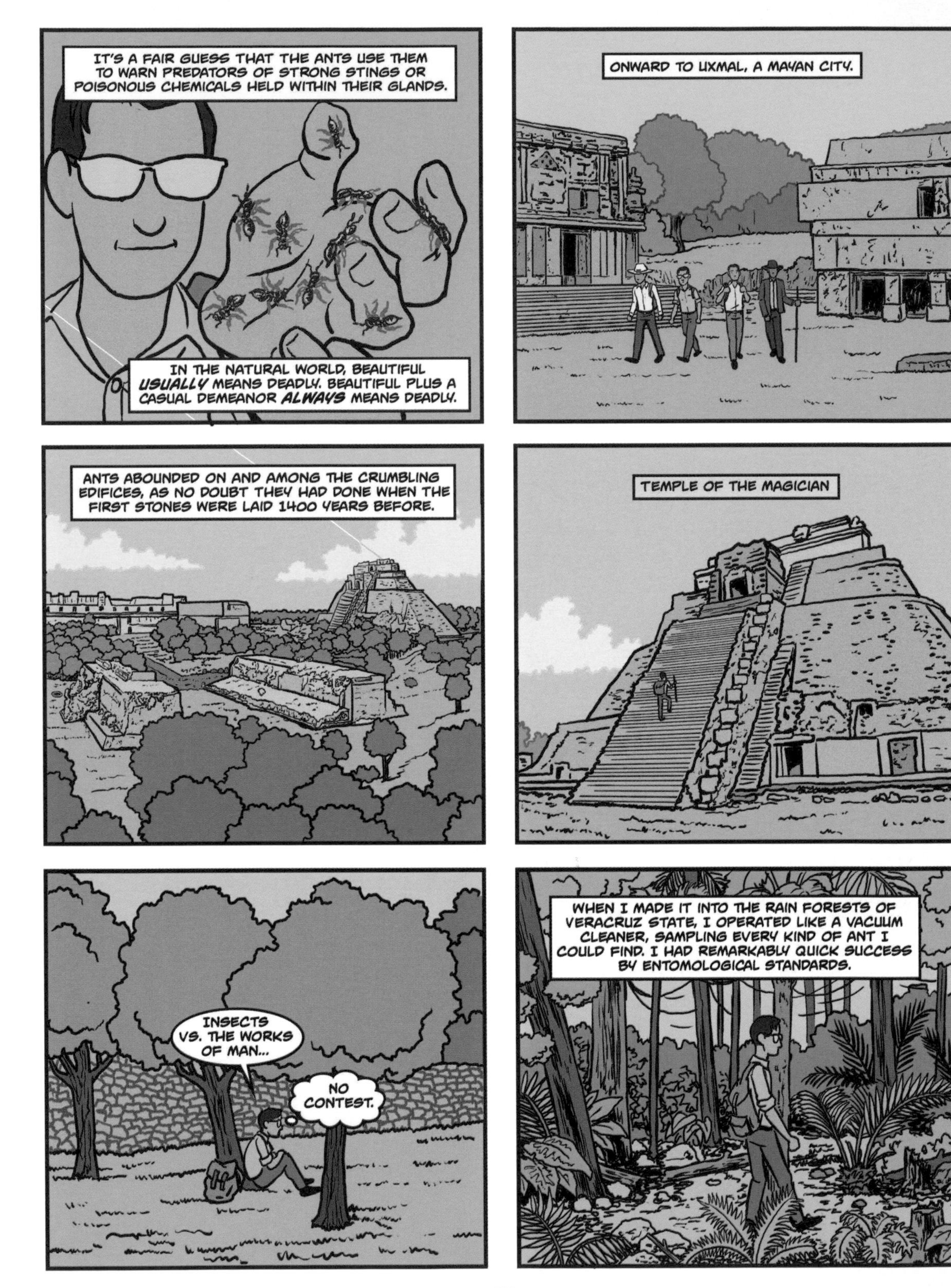

IT'S A FAIR GUESS THAT THE ANTS USE THEM TO WARN PREDATORS OF STRONG STINGS OR POISONOUS CHEMICALS HELD WITHIN THEIR GLANDS.
IN THE NATURAL WORLD, BEAUTIFUL *USUALLY* MEANS DEADLY. BEAUTIFUL PLUS A CASUAL DEMEANOR *ALWAYS* MEANS DEADLY.
ONWARD TO UXMAL, A MAYAN CITY.
ANTS ABOUNDED ON AND AMONG THE CRUMBLING EDIFICES, AS NO DOUBT THEY HAD DONE WHEN THE FIRST STONES WERE LAID 1400 YEARS BEFORE.
TEMPLE OF THE MAGICIAN
INSECTS VS. THE WORKS OF MAN...
NO CONTEST.
WHEN I MADE IT INTO THE RAIN FORESTS OF VERACRUZ STATE, I OPERATED LIKE A VACUUM CLEANER, SAMPLING EVERY KIND OF ANT I COULD FIND. I HAD REMARKABLY QUICK SUCCESS BY ENTOMOLOGICAL STANDARDS.

BELONOPELTA
I CAPTURED COLONIES OF TWO GENERA THAT HAD NEVER BEEN STUDIED BEFORE, AND RECORDED MY OBSERVATIONS ON THEIR SOCIAL ORGANIZATION AND PREDATORY BEHAVIOR FOR LATER PUBLICATION.
AS I PREPARED TO LEAVE THE VERACRUZ COAST TWO WEEKS LATER, MY ATTENTION WAS DRAWN TO PICO DE ORIZABA, THE GREAT VOLCANIC MOUNTAIN JUST NORTH OF THE CITY OF ORIZABA.
5,747 M/18,855 FT.

I WAS DRAWN NOT JUST BY THE AMAZING SIGHT BUT ALSO BY THE VERY CONCEPT OF ORIZABA.

I THOUGHT OF THE MOUNTAIN AS AN ISLAND.

IT WAS ISOLATED FROM THE PLATEAU, YET I BELIEVED THAT A LONE CLIMBER COULD TRAVEL IN ONE RELATIVELY SHORT, STRAIGHT PASS FROM TROPICAL FOREST TO COLD TEMPERATE FOREST...

AND FINALLY INTO THE TREELESS ARCTIC SCREE JUST BELOW THE SUMMIT.

THE COOLER HABITATS CONSTITUTED THE ISLAND. THE SURROUNDING TROPICAL AND SUBTROPICAL LOWLANDS WERE THE SEA.

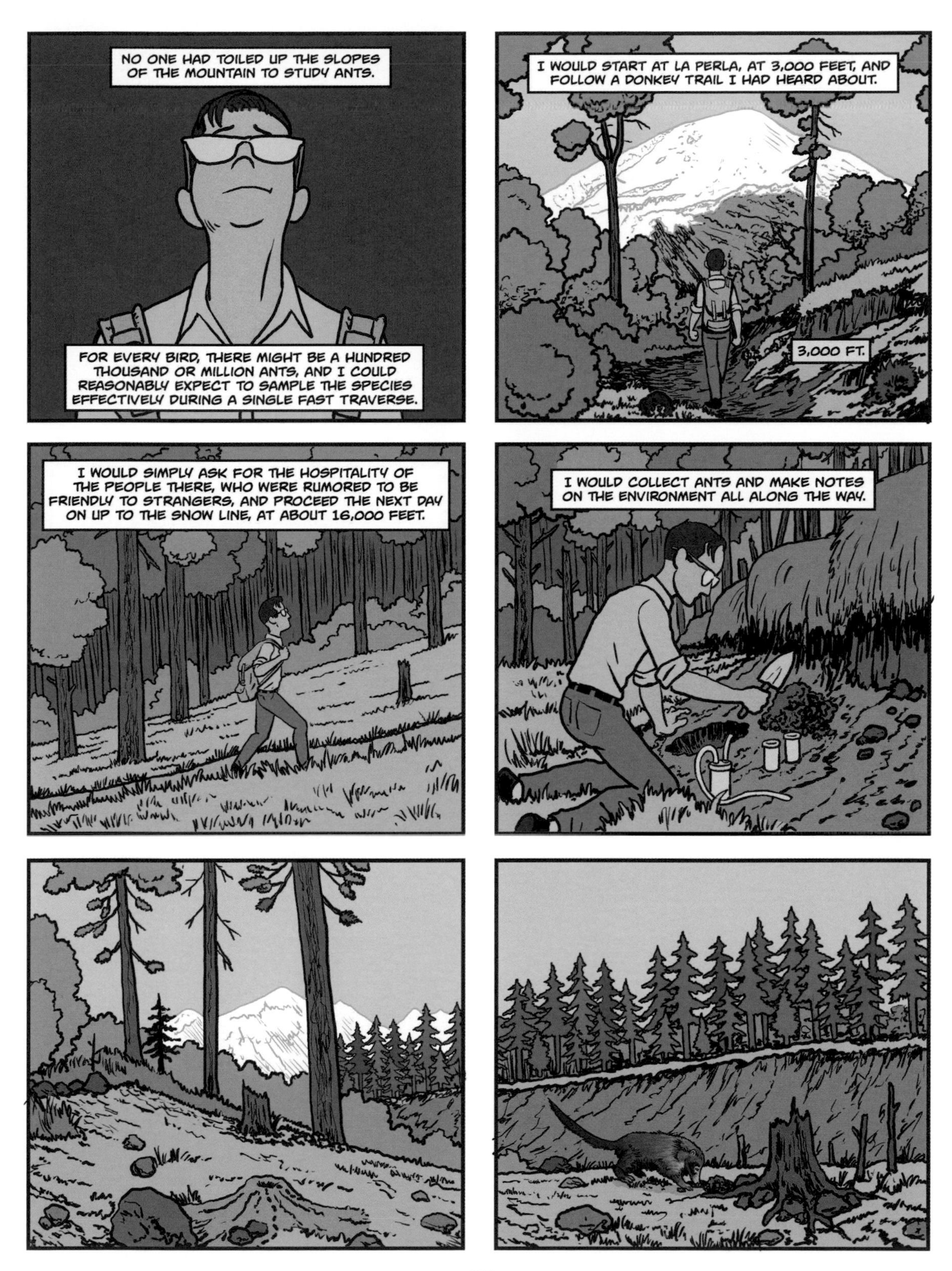
NO ONE HAD TOILED UP THE SLOPES OF THE MOUNTAIN TO STUDY ANTS.
FOR EVERY BIRD, THERE MIGHT BE A HUNDRED THOUSAND OR MILLION ANTS, AND I COULD REASONABLY EXPECT TO SAMPLE THE SPECIES EFFECTIVELY DURING A SINGLE FAST TRAVERSE.
I WOULD START AT LA PERLA, AT 3,000 FEET, AND FOLLOW A DONKEY TRAIL I HAD HEARD ABOUT.
3,000 FT.
I WOULD SIMPLY ASK FOR THE HOSPITALITY OF THE PEOPLE THERE, WHO WERE RUMORED TO BE FRIENDLY TO STRANGERS, AND PROCEED THE NEXT DAY ON UP TO THE SNOW LINE, AT ABOUT 16,000 FEET.
I WOULD COLLECT ANTS AND MAKE NOTES ON THE ENVIRONMENT ALL ALONG THE WAY.

I WAS A FOOL, OF COURSE.
ALONE, ON FOOT, UP A HIGH MOUNTAIN, PHRASEBOOK SPANISH
NO MAP
WELL, IT'S A PATH, ANYWAY.
5,500 FT.
THE ANTS IN THIS TRANSITION BELT, WHICH COMPOSED NOTHING LESS THAN THE PASSAGE FROM THE NEOTROPICAL TO THE NEARCTIC REGIONS, WERE A MIX OF TROPICAL AND TEMPERATE SPECIES.
TWO OF THE *FORMICA* I FOUND LATER PROVED NEW TO SCIENCE.
8,000 FT.
WHEN I ARRIVED AT RANCHO SOMECLA, I WAS CLOSE TO EXHAUSTION.
11,000 FT.

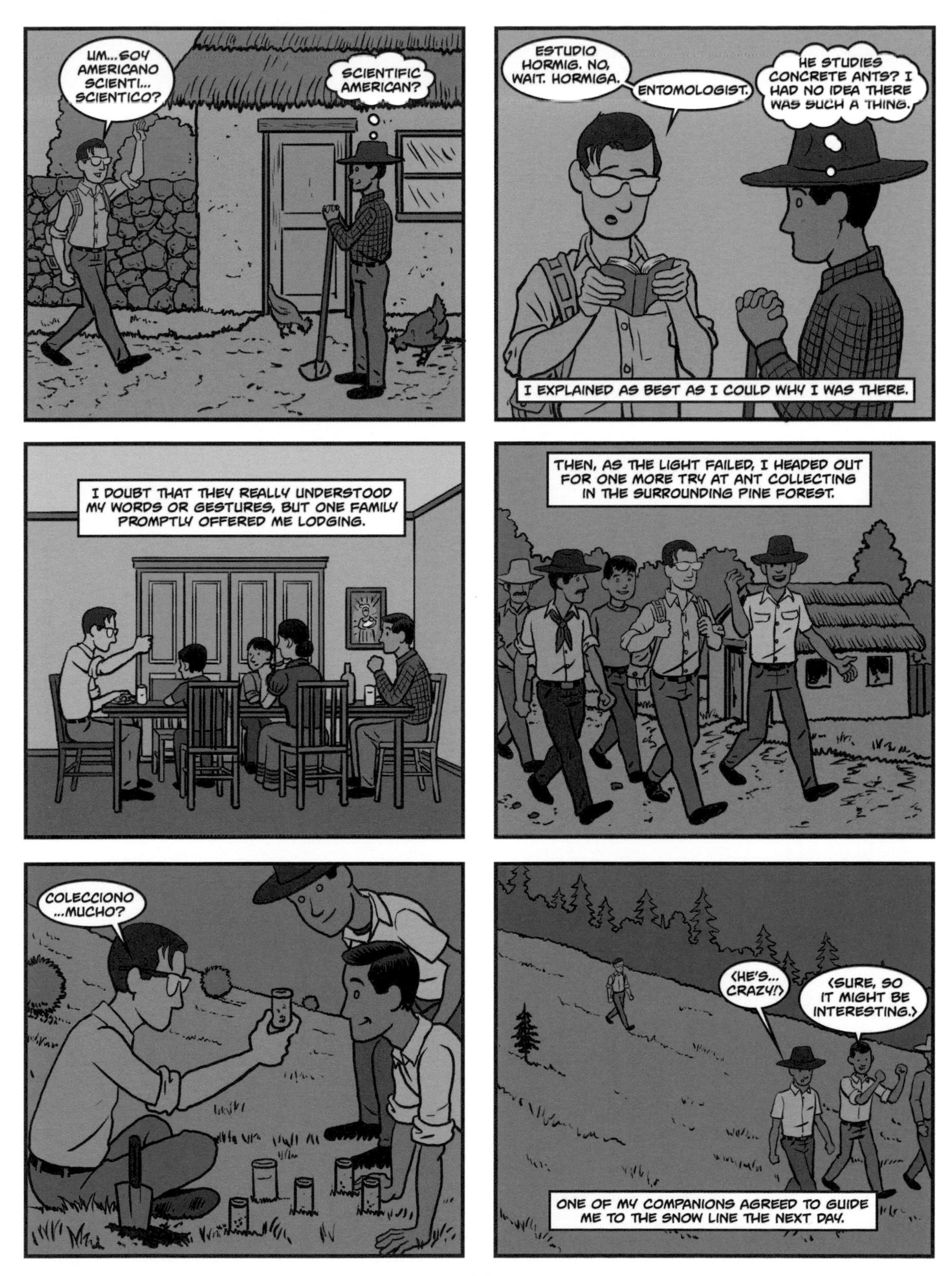
UM...SOY AMERICANO SCIENTI... SCIENTICO?
SCIENTIFIC AMERICAN?
ESTUDIO HORMIG. NO, WAIT. HORMIGA.
ENTOMOLOGIST.
HE STUDIES CONCRETE ANTS? I HAD NO IDEA THERE WAS SUCH A THING.
I EXPLAINED AS BEST AS I COULD WHY I WAS THERE.
I DOUBT THAT THEY REALLY UNDERSTOOD MY WORDS OR GESTURES, BUT ONE FAMILY PROMPTLY OFFERED ME LODGING.
THEN, AS THE LIGHT FAILED, I HEADED OUT FOR ONE MORE TRY AT ANT COLLECTING IN THE SURROUNDING PINE FOREST.
COLECCIONO ...MUCHO?
‹HE'S... CRAZY!›
‹SURE, SO IT MIGHT BE INTERESTING.›
ONE OF MY COMPANIONS AGREED TO GUIDE ME TO THE SNOW LINE THE NEXT DAY.

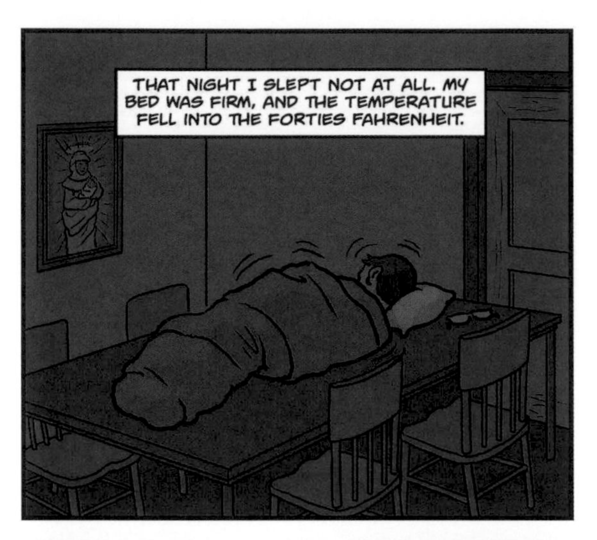
THAT NIGHT I SLEPT NOT AT ALL. MY BED WAS FIRM, AND THE TEMPERATURE FELL INTO THE FORTIES FAHRENHEIT.

THIS WOULD BE A WONDERFUL PLACE TO LIVE, IF YOU BROUGHT A LOT OF BLANKETS.

WE LEFT AT DAWN.

MY EXCITEMENT GREW WHEN WE ENTERED THE OPEN CLOUD FOREST, BUT...
SOMEWHERE BETWEEN 12,000 AND 13,000 FT.

BUENO.
NO <GASP> MÁS.
¿ABAJO?

AH...BAH...HO?
DOWN. YES.
I MEAN, SÍ, POR FAVOR.
...BUT THE AIR WAS TOO THIN FOR SOMEONE WHO HAD BEEN LIVING AT SEA LEVEL, AND I WAS GASPING FOR BREATH.
IN ANY CASE, ANTS HAD BECOME VERY SCARCE, EVEN IN THE CLEARINGS WARMED BY THE MORNING SUN. I SEARCHED FOR AN HOUR BEFORE FINDING ONE COLONY NESTING BENEATH A WOOD CHIP.
AT RANCHO SOMECLA, I SHOOK HANDS WITH MY GUIDE AND HEADED ALONE DOWN THE TRAIL TO LA PERLA, MOVING RAPIDLY NOW, THEN TO MY HOTEL IN ORIZABA.
THERE, THE SATED ADVENTURER, I SLEPT FOR TWELVE HOURS.
ZZZZ

PART TWO
Storyteller

IF YOU'RE A STORYTELLER,
FIND A GOOD STORY AND TELL IT.

[HOWARD HAWKS, FILMMAKER]

LESS THAN A YEAR LATER, ON A COLD MARCH DAY IN 1954, PHILIP DARLINGTON CALLED ME TO HIS OFFICE.
HE WAS CURATOR IN ENTOMOLOGY AT HARVARD'S MUSEUM OF COMPARATIVE ZOOLOGY.
HOW WOULD YOU LIKE TO GO TO NEW GUINEA?

YES.
THE SOCIETY OF FELLOWS AND MUSEUM OF COMPARATIVE ZOOLOGY HAS AGREED TO COVER YOUR...
YES.
EXPENSES... FOR AN EXTENDED VISIT.
YES.

NO SPECIALIST HAD COLLECTED ANTS IN THAT FABULOUSLY RICH AND STILL MOSTLY UNEXPLORED FAUNA...
WHO KNOWS WHAT YOU'LL FIND? AND IF YOU PICK UP SOME BEETLES FOR ME, THAT WOULD BE ALL RIGHT TOO.

AFTER REMINDING ME OF HIS COLLECTING ADVICE...
LIKE I TOLD YOU BEFORE, DON'T STAY ON THE TRAILS. WALK IN A STRAIGHT LINE THROUGH THE FOREST. TRY TO GO OVER ANY BARRIER YOU MEET.
...ADVICE I HAD FOLLOWED ALL TOO WELL ON ORIZABA, HE SENT ME ON MY WAY.

GO, WHILE YOU'RE STILL FOOTLOOSE AND FANCY-FREE.
NOW TELL ME, HOW DID YOU FIND CUBAN COFFEE? DID YOU DRINK IT THE WAY I TOLD YOU?

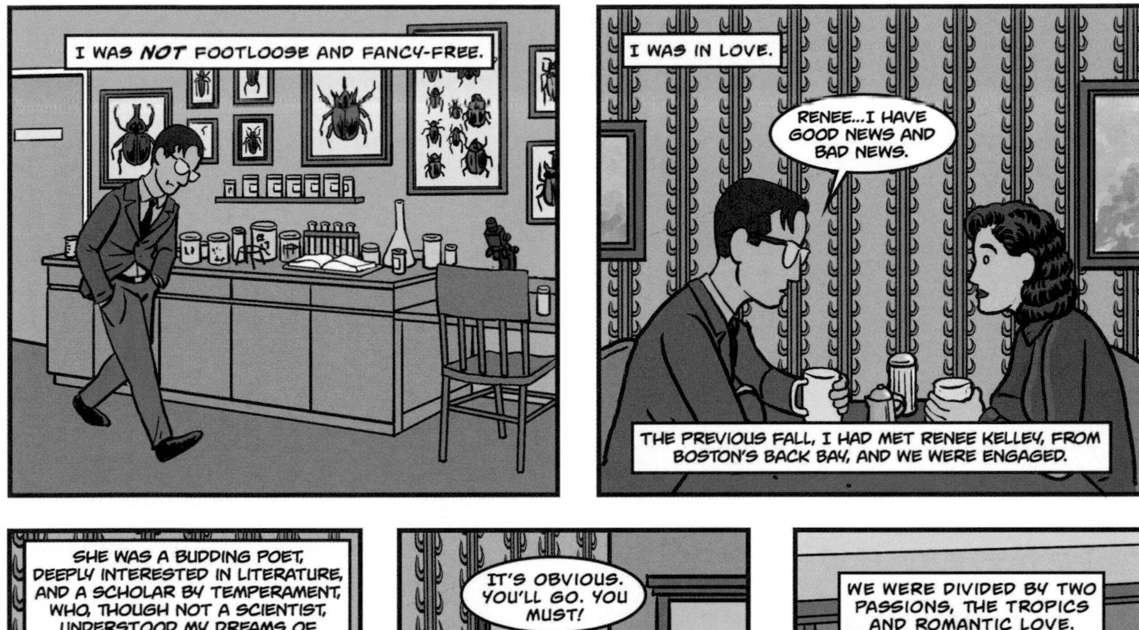
I WAS NOT FOOTLOOSE AND FANCY-FREE.
I WAS IN LOVE.
RENEE...I HAVE GOOD NEWS AND BAD NEWS.
THE PREVIOUS FALL, I HAD MET RENEE KELLEY, FROM BOSTON'S BACK BAY, AND WE WERE ENGAGED.

SHE WAS A BUDDING POET, DEEPLY INTERESTED IN LITERATURE, AND A SCHOLAR BY TEMPERAMENT, WHO, THOUGH NOT A SCIENTIST, UNDERSTOOD MY DREAMS OF PURSUITS IN FARAWAY PLACES.
I JUST DON'T KNOW WHAT I... WE SHOULD DO. WHAT SHOULD WE DO?

IT'S OBVIOUS. YOU'LL GO. YOU MUST!

WE WERE DIVIDED BY TWO PASSIONS, THE TROPICS AND ROMANTIC LOVE.

TEMPORARILY DIVIDED.

WE WROTE TO EACH OTHER DAILY AND AT LENGTH, ACCUMULATING A TOTAL OF SOME SIX HUNDRED DIARY-LIKE LETTERS.

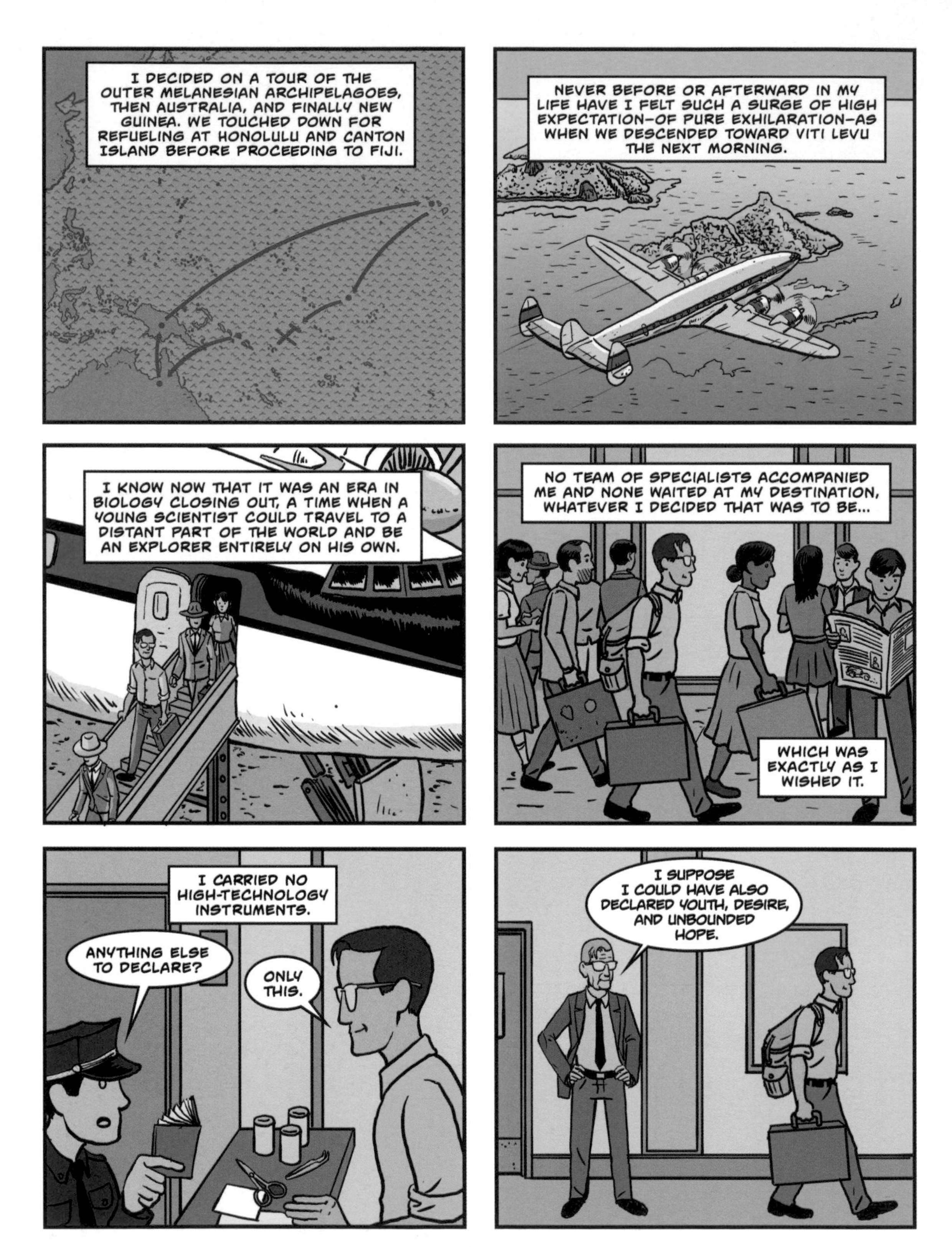
I DECIDED ON A TOUR OF THE OUTER MELANESIAN ARCHIPELAGOES, THEN AUSTRALIA, AND FINALLY NEW GUINEA. WE TOUCHED DOWN FOR REFUELING AT HONOLULU AND CANTON ISLAND BEFORE PROCEEDING TO FIJI.
NEVER BEFORE OR AFTERWARD IN MY LIFE HAVE I FELT SUCH A SURGE OF HIGH EXPECTATION—OF PURE EXHILARATION—AS WHEN WE DESCENDED TOWARD VITI LEVU THE NEXT MORNING.
I KNOW NOW THAT IT WAS AN ERA IN BIOLOGY CLOSING OUT, A TIME WHEN A YOUNG SCIENTIST COULD TRAVEL TO A DISTANT PART OF THE WORLD AND BE AN EXPLORER ENTIRELY ON HIS OWN.
NO TEAM OF SPECIALISTS ACCOMPANIED ME AND NONE WAITED AT MY DESTINATION, WHATEVER I DECIDED THAT WAS TO BE...
WHICH WAS EXACTLY AS I WISHED IT.
I CARRIED NO HIGH-TECHNOLOGY INSTRUMENTS.
ANYTHING ELSE TO DECLARE?
ONLY THIS.
I SUPPOSE I COULD HAVE ALSO DECLARED YOUTH, DESIRE, AND UNBOUNDED HOPE.

THE SOUTH PACIFIC IS A GALAXY OF THOUSANDS OF ISLANDS, SPREAD OUT IN CONFIGURATIONS THAT HAVE SERVED MANY OF THE KEY ADVANCES IN EVOLUTIONARY BIOLOGY.
DARWIN CONCEIVED OF EVOLUTION BY NATURAL SELECTION FROM WHAT HE LEARNED IN THE GALAPAGOS ISLANDS.
ALFRED RUSSEL WALLACE HAD THE SAME IDEA AFTER STUDYING BUTTERFLIES IN THE OLD MALAY ARCHIPELAGO.
BY FACTORING IN THE AGE OF AN ISLAND AND THE ORIGIN OF ITS IMMIGRANTS, BIOLOGISTS CAN RECONSTRUCT THE EVOLUTION OF THE PLANTS AND ANIMALS THERE MORE EASILY THAN ON CONTINENTS.
THE SIMPLICITY OF ISLANDS MAKES THEM THE BEST OF ALL NATURAL LABORATORIES.
UNLIKE EXPERIMENTAL BIOLOGISTS, EVOLUTIONARY BIOLOGISTS ALREADY HAVE AN ABUNDANCE OF ANSWERS FROM WHICH TO PICK AND CHOOSE. WHAT THEY MOST NEED ARE THE RIGHT QUESTIONS.
THE MOST IMPORTANT EVOLUTIONARY BIOLOGISTS ARE THOSE WHO INVENT THE MOST IMPORTANT QUESTIONS.
IF THEY ARE ALSO NATURALISTS, THEY GO INTO THE FIELD WITH OPEN EYES AND MINDS, LOOKING IN ALL DIRECTIONS FOR THE BIG QUESTIONS, FOR THE MAIN CHANCE.

THEY MUST KNOW ONE OR TWO GROUPS OF PLANTS OR ANIMALS WELL ENOUGH TO MAKE THEM ACTORS IN THE THEATER OF THEIR VISION.

THEY LOOK FOR THE BEST STORIES NATURE HAS TO TELL US, BECAUSE THEY ARE, ABOVE ALL, STORYTELLERS.

NADALA, VITI LEVU, DECEMBER 1954.

MY DARLING RENEE,
FIJI IS, IN ONE DREADFUL SENSE, CUBA AND MEXICO ALL OVER AGAIN: THE NATIVE BIOTA HAS BEEN DRIVEN BACK TO SCATTERED AND NEARLY INACCESSIBLE ENCLAVES.

I'VE LEARNED ANOTHER MELANCHOLY FACT ABOUT CONSERVATION. IN A SMALL PATCH OF WHAT APPEARED TO BE NATURAL FOREST I FOUND ONLY INVASIVE ANT SPECIES.

ON ISLANDS HARBORING NATIVE SPECIES OF LIMITED DIVERSITY, THE ECOSYSTEMS ARE VULNERABLE TO INVASION BY ALIENS EVEN IF LEFT PHYSICALLY INTACT.

MUCH OF THE PACIFIC FAUNA HAS GONE UNDER IN THE PATH OF PIGS, GOATS, AND OTHER HIGHLY COMPETITIVE FORMS INTRO-DUCED BY HUMAN COMMERCE. STRANGERS HAVE SAVAGED THE ISLANDS OF THE WORLD.

THE ANT FAUNA HERE IS ALREADY REASONABLY WELL KNOWN, THANKS TO WILLIAM MANN'S LENGTHY RESIDENCE.

"SO, I WILL NOT LINGER IN FIJI."

MOUNT MOU, NEW CALEDONIA, DECEMBER 1954.
MY DARLING RENEE:
HERE, NOT JUST THE ANTS BUT EVERYTHING I SEE, EVERY SPECIES OF PLANT AND ANIMAL, IS NEW TO ME. THESE CREATURES ARE A FULLY ALIEN BIOTA, AND IT IS TIME TO CONFESS...
I AM A NEOPHILE, AN INORDINATE LOVER OF THE NEW, OF DIVERSITY FOR ITS OWN SAKE. IN SUCH A PLACE, EVERYTHING IS A SURPRISE, AND I COULD MAKE A DISCOVERY OF SCIENTIFIC VALUE ANYTIME I WISHED.
MY ARCHETYPAL DREAM IS CLEAR TO ME...
TAKE ME, LORD, TO AN UNEXPLORED PLANET TEEMING WITH NEW LIFE FORMS. PUT ME AT THE EDGE OF VIRGIN SWAMPLAND DOTTED WITH HUMMOCKS OF HIGH GROUND.
LET ME SAUNTER ACROSS IT AND UP THE NEAREST MOUNTAIN RIDGE, AND CROSS OVER TO THE FAR SLOPE IN SEARCH OF MORE DISTANT SWAMPS, GRASSLANDS, AND RANGES.
LET ME BE THE CAROLUS LINNAEUS OF THIS WORLD, BEARING NO MORE THAN SPECIMEN BOXES, BOTANICAL CANISTER, HAND LENS, NOTEBOOKS, BUT ALLOWED NOT YEARS BUT CENTURIES OF TIME.
LET ME GO ALONE, AT LEAST FOR WHILE, AND I WILL REPORT TO YOU AND LOVED ONES AT INTERVALS, AND I WILL PUBLISH REPORTS ON MY DISCOVERIES FOR COLLEAGUES.
ALONE?
FOR IF IT WAS YOU WHO GAVE ME THIS SPIRIT, THEN DEVISE THE APPROPRIATE REWARD FOR ITS VIRTUOUS USE.

CIU, NEAR MOUNT CANALA, NEW CALEDONIA, DECEMBER 1954.
THE FÉRÉ TRACT IS NOT A TRUE RAIN FOREST IN THE FAMILIAR, AMAZONIAN SENSE. IT COMPRISES ONLY TWO STORIES OF TREES, WITH THE UPPER CANOPY TWENTY METERS HIGH AND BROKEN IN ENOUGH PLACES TO LET SUNSHINE FALL IN LARGE, RADIANT PATCHES ON THE FOREST FLOOR.
THE HABITAT IS IDEAL FOR ANTS. IT ABOUNDS WITH PURE NEW CALEDONIAN SPECIES, MANY NEW TO SCIENCE. I AM STRUCK BY THE PREVALENCE OF RED-AND-BLACK COLORATION AMONG THE WORKERS FORAGING ABOVE GROUND.
AT CHAPEAU GENDARME, NEAR NOUMEA, THE SAME SPECIES WERE PREDOMINANTLY YELLOW. WHAT IS THE MEANING OF THIS LOCAL COLOR CODE? PERHAPS IT IS JUST COINCIDENCE. BUT I SUSPECT MIMICRY. MY GUESS IS THAT ONE TO SEVERAL OF THE SPECIES ARE POISONOUS, AS I SUPPOSED TO BE TRUE OF THE METALLESCENT ANTS OF CUBA.
IN THEORY, IT PAYS FOR ALL LOCAL POISONOUS SPECIES TO EVOLVE THE SAME COLOR, FORMING A CONSORTIUM AMONG THE ADVERTISERS. IT ALSO PAYS FOR HARMLESS, TASTY SPECIES TO ACQUIRE THE SAME APPEARANCE AND ENJOY A FREE RIDE ON THE REPELLANT FORMS THEY IMITATE.
I HAVE NEITHER THE MEANS NOR THE TIME, HOWEVER, TO TEST EITHER HYPOTHESIS.

MOUNT RAGGED, WESTERN AUSTRALIA, JANUARY-FEBRUARY 1955.
MY DARLING:
I'M NOW IN AUSTRALIA FOR A POTENTIALLY EVEN MORE IMPORTANT EXCURSION.
I TOOK THE WEEKLY QANTAS FLYING BOAT BACK TO NOUMEA, THEN TO SYDNEY AND, AFTER A BRIEF STAY IN THE CITY AND A COLLECTING EXCURSION INTO THE SURROUNDING COUNTRYSIDE, FLEW ON TO KALGOORLIE.
FROM THIS INLAND CENTER OF WESTERN AUSTRALIA'S SHEEP COUNTRY, I PROCEEDED SOUTH BY RAIL TO NORSEMAN FOR A ROUND OF ANT COLLECTING.
AT A LOCAL BAR, I FELL IN WITH A GROUP OF CONSTRUCTION WORKERS WHO INVITED ME TO COLLECT ANTS OUT AT THEIR WORKPLACE IN THE NEARBY EUCALYPTUS SCRUB.
A FULL DAY IN THE BUSH COMPLETELY DEHYDRATED ME—TWO MONTHS IN THE HUMID TROPICS HAVE RENDERED MY SYSTEM UNABLE TO HANDLE EVAPORATION IN SUCH A HOT, SEMIDESERT ENVIRONMENT.
WHEN WE ARRIVED BACK AT THE BAR LATE THAT AFTERNOON, I CHUGALUGGED FOUR BEERS IN A ROW. MY HOSTS, THEMSELVES HEAVY CONSUMERS IN A COUNTRY KNOWN FOR OLYMPIC-CLASS BEER DRINKING, WERE IMPRESSED.
I'M NOT SURE "IMPRESSED" IS THE RIGHT WORD HERE.
NORMALLY A ONE-BEER-MAXIMUM OCCASIONAL DRINKER, SO AM I.

I THEN WENT FARTHER SOUTH TO ESPERANCE, AN ISOLATED COASTAL TOWN JUST WEST OF THE GREAT AUSTRALIAN BIGHT.
HERE I WAS JOINED BY CARYL HASKINS, A FELLOW ENTOMOLOGIST AND THE NEWLY APPOINTED PRESIDENT OF THE CARNEGIE INSTITUTION OF WASHINGTON.
A HUNDRED KILOMETERS TO THE EAST, OUT ACROSS THE SANDPLAIN HEATH, LIVED THE GRAIL OF ANT STUDIES—*NOTHOMYRMECIA MACROPS*, THE MOST PRIMITIVE KNOWN ANT, A LOST SPECIES SINCE ITS DISCOVERY.
"IT IS QUITE POSSIBLY THE KEY TO THE ORIGIN OF SOCIAL LIFE IN ANTS. WE MEANT TO REDISCOVER THE SPECIES AND BE THE FIRST TO STUDY IT IN LIFE. THE WHOLE STORY BEGAN ON DECEMBER 7, 1931..."
...WHEN A HOLIDAY PARTY SET OUT FROM BALLADONIA, A SHEEP RANCH AND BEER STOP ON THE CROSS-AUSTRALIA HIGHWAY NORTHEAST OF ESPERANCE.

THEY TRAVELED LEISURELY FOR 175 KILOMETERS SOUTHWARD.
AFTER PASSING MOUNT RAGGED, THEY STOPPED FOR A FEW DAYS AT THE ABANDONED THOMAS RIVER STATION ON THE COAST.
THE HABITAT IS BOTANICALLY ONE OF THE RICHEST IN THE WORLD, AND MRS. AMY CROCKER, A NATURALIST AND ARTIST AT BALLADONIA, HAD ASKED MEMBERS OF THE PARTY TO COLLECT INSECTS ALONG THE WAY.
THE SPECIMENS, INCLUDING TWO LARGE, ODDLY SHAPED YELLOW ANTS, WERE TURNED OVER TO THE NATIONAL MUSEUM OF VICTORIA IN MELBOURNE.
THERE, THE ANTS WERE DESCRIBED BY THE ENTOMOLOGIST JOHN CLARK IN 1934.
HMM...CLEARLY A NEW GENUS AND SPECIES.
QUITE PRIMITIVE, AREN'T YOU?
NOTHOMYRMECIA MACROPS

WE FOUND THE THOMAS RIVER TO BE A DRY BED—AN ARROYO—IN A BASIN DEPRESSED TWENTY-FIVE TO THIRTY METERS BELOW THE LEVEL OF THE SANDPLAIN.

NOTHOMYRMECIA COULD HAVE BEEN ANYWHERE IN SUCH A VARIED ENVIRONMENT.

I WAS EXCITED AND TENSE, KNOWING THAT WE MIGHT FIND SCIENTIFIC GOLD WITH A SINGLE GLANCE TO THE GROUND. HASKINS AND I SET TO WORK IMMEDIATELY, EACH HOPING TO BE THE LUCKY DISCOVERER.

...

THAT NIGHT, ARMED WITH FLASHLIGHTS AND NET, WE WALKED BACK OUT ONTO THE SANDPLAIN, AND THIS TIME LOST OUR WAY. RATHER THAN RISK WANDERING FARTHER FROM CAMP IN A DANGEROUS DESERT-LIKE ENVIRONMENT, WE SETTLED DOWN TO WAIT FOR DAYBREAK.

TO MY SURPRISE, CARYL FOUND A FOOTBALL-SIZED STONE, PULLED AND ROCKED IT AS THOUGH POSITIONING A PILLOW, LAY ON HIS BACK ON THE GROUND, AND FELL ASLEEP.

I WAS TOO KEYED UP TO ATTEMPT THE SAME FEAT AND SPENT THE REST OF THE NIGHT SEARCHING FOR THE ANT IN THE IMMEDIATE VICINITY. HOW MARVELOUS IT WOULD BE, I THOUGHT, IF I COULD HAND CARYL A SPECIMEN WHEN HE AWOKE!

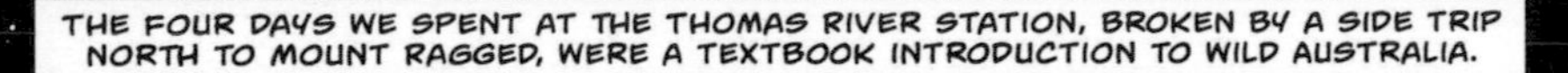

RESEARCH PROGRESS WAS RAPID AND SATISFYING, AT LEAST BY ORDINARY STANDARDS OF FIELD BIOLOGY.

WE DISCOVERED NEW SPECIES, IN THE COURSE OF WHICH WE ALSO DEFINED AN ENTIRE ECOLOGICAL GUILD OF SANDPLAIN ANTS SPECIALIZED FOR FORAGING ON THE LOW VEGETATION AT NIGHT.

LARGE-EYED AND LIGHT-COLORED, THEY REPRESENT MEMBERS OF THE GENERA *CAMPONOTUS*, *COLOBOSTRUMA*, AND *IRIDOMYRMEX* THAT HAVE EVIDENTLY CONVERGED IN EVOLUTION TO FILL THIS ARID NICHE.

BECAUSE *NOTHOMYRMECIA* IS ALSO LARGE-EYED AND PALE, WE REASONED THAT IT WAS A MEMBER OF THE GUILD, AND SO WE CONCENTRATED OUR EFFORTS ON THE SANDPLAIN.

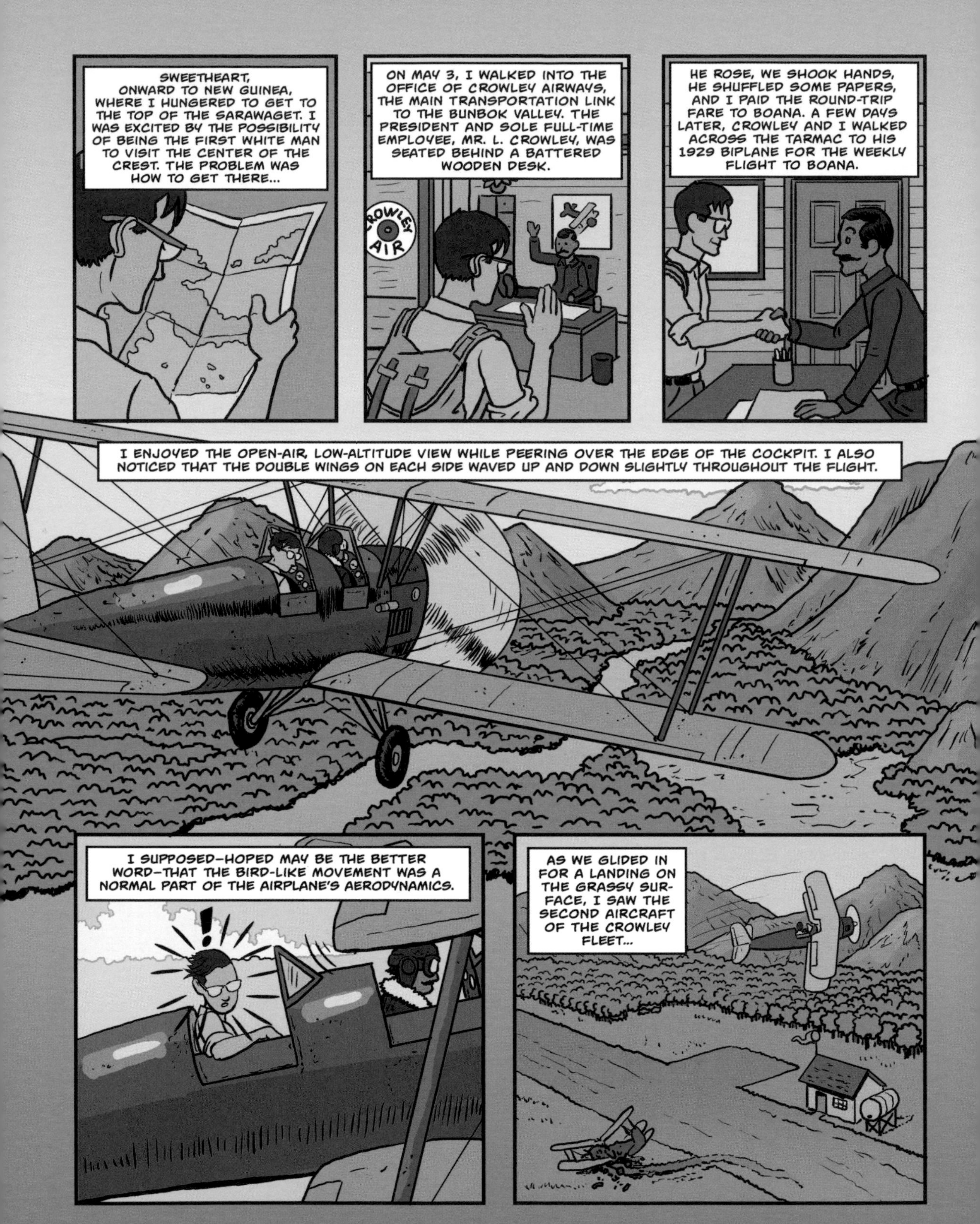
SWEETHEART, ONWARD TO NEW GUINEA, WHERE I HUNGERED TO GET TO THE TOP OF THE SARAWAGET. I WAS EXCITED BY THE POSSIBILITY OF BEING THE FIRST WHITE MAN TO VISIT THE CENTER OF THE CREST. THE PROBLEM WAS HOW TO GET THERE...
ON MAY 3, I WALKED INTO THE OFFICE OF CROWLEY AIRWAYS, THE MAIN TRANSPORTATION LINK TO THE BUNBOK VALLEY. THE PRESIDENT AND SOLE FULL-TIME EMPLOYEE, MR. L. CROWLEY, WAS SEATED BEHIND A BATTERED WOODEN DESK.
CROWLEY AIR
HE ROSE, WE SHOOK HANDS, HE SHUFFLED SOME PAPERS, AND I PAID THE ROUND-TRIP FARE TO BOANA. A FEW DAYS LATER, CROWLEY AND I WALKED ACROSS THE TARMAC TO HIS 1929 BIPLANE FOR THE WEEKLY FLIGHT TO BOANA.
I ENJOYED THE OPEN-AIR, LOW-ALTITUDE VIEW WHILE PEERING OVER THE EDGE OF THE COCKPIT. I ALSO NOTICED THAT THE DOUBLE WINGS ON EACH SIDE WAVED UP AND DOWN SLIGHTLY THROUGHOUT THE FLIGHT.
I SUPPOSED—HOPED MAY BE THE BETTER WORD—THAT THE BIRD-LIKE MOVEMENT WAS A NORMAL PART OF THE AIRPLANE'S AERODYNAMICS.
!
AS WE GLIDED IN FOR A LANDING ON THE GRASSY SURFACE, I SAW THE SECOND AIRCRAFT OF THE CROWLEY FLEET...

IN THE LATE MORNING, THE SARAWAGET CREST SEEMED CLOSE TO BOANA, A DAY'S WALK PERHAPS, AND I ACHED TO GET STARTED.
IT WAS, IN FACT, FIVE DAYS' WALK AWAY...SELDOM WERE WE ABLE TO TRAVEL FOR MORE THAN A HUNDRED METERS IN A STRAIGHT LINE ON LEVEL GROUND.
WE WOVE, STUMBLED, WADED, CLIMBED, AND SOMETIMES JUST CRAWLED OUR WAY, FOLLOWING STREAM BANKS, TRACING ANIMAL TRAILS UP AND ALONG THE CRESTS OF RIDGES, DOWN INTO STREAM VALLEYS AND THEN UP AGAIN.
WE WERE ALMOST CONSTANTLY WET FROM INTERMITTENT RAIN, WHICH SET IN PREDICTABLY BY EARLY AFTERNOON AND CONTINUED INTO THE EARLY EVENING.
LAND LEECHES WERE EVERYWHERE, AND PERIODICALLY WE STOPPED TO TAKE THEM OFF OUR LEGS AND FEET. TO SIT ON A MUDDY STREAM BANK IN NEAR EXHAUSTION, PULL OFF BOOTS AND PEEL DOWN SOCKS, AND BURN FREE A HALF-DOZEN ENGORGED LEECHES, THEN WATCH BLOOD TRICKLE DOWN FROM THE BITE WOUNDS—THAT IS AN EXPERIENCE BEST SAVORED AFTER A FEW YEARS HAVE PASSED.
I WAS AFRAID MOST OF ALL, AT TIMES, OF THE INEXPRESSIBLE UNKNOWN.
WOULD I FAIL FROM PHYSICAL INCAPACITY OR LACK OF WILL? WOULD I HAVE TO TURN BACK AS I DID IN MEXICO, SHORT OF THE ORIZABA SNOWFIELD?
WHY HAD I COME HERE, ANYWAY, EXCEPT TO BE ABLE TO SAY I WAS THE FIRST WHITE MAN TO CLIMB THE CENTRAL SARAWAGET?

THAT PRIDEFUL GOAL WAS PART OF THE TRUTH, BUT I WAS LOOKING FOR SOMETHING MORE. I WANTED TO BE THE FIRST NATURALIST TO WALK THE SARAWAGET CREST AND COLLECT ANIMALS THERE.
AND I WANTED RELEASE FROM THE MOUNTAIN-PEAK COMPULSION THAT GRIPPED ME.
EARLY ON THE FIFTH DAY, BEFORE THE INEVITABLE CLOUDS DESCENDED AND COLD RAIN BEGAN TO FALL, WE WALKED THE FINAL TWO HOURS TO THE SUMMIT.
EDWARD OSBORNE
NE WILSON
AS WE FAST-WALKED AND SLID OUR WAY DOWN THE RIDGES AND INTO THE UPPER BUNBOK VALLEY, SOMETHING RESISTANT AND TROUBLING FINALLY BROKE AND RECEDED INSIDE ME.
THE SARAWAGET, COLD AND DAUNTING, HAD PROVED A TEST OF MY WILL SEVERE ENOUGH TO BE SATISFYING. I HAD REACHED THE EDGE OF THE WORLD I WANTED, AND KNEW MYSELF BETTER AS A RESULT.
BY PASSING FROM THE SEA TO ITS PEAKS, I HAD FINALLY ENCOMPASSED THE SERIOUS TROPICS OF MY DREAMS, AND I COULD GO HOME.

CEYLON, SUMMER 1955.
~~SWEETHEART:~~
I CONTINUE WESTWARD AND AM NOW IN CEYLON, THE ISLAND "PEARL OF ASIA" THAT HANGS LIKE A TEARDROP FROM THE TIP OF INDIA—WITH STOPS IN QUEENSLAND AND PERTH ALONG THE WAY.
I TRAVELED INLAND FROM THE PORT OF COLOMBO TO SEARCH FOR ONE OF THE RAREST ANTS IN THE WORLD. *ANEURETUS SIMONI* IS THE APPARENT EVOLUTIONARY LINK BETWEEN TWO OF THE GREAT WORLDWIDE GROUPS OF ANTS, THE *MYRMICINAE* AND *DOLICHODERINAE*.
NOW ONLY ONE SPECIES REMAINS, THE ENDANGERED *ANEURETUS SIMONI*. I BEGAN MY QUEST IN THE BOTANIC GARDENS AT PERADENIYA, WHERE THE ONLY SPECIMENS IN MUSEUMS—ALL BELONGING TO THE WORKER CASTE—WERE COLLECTED AROUND 1890.
FOR THREE DAYS, I WORKED WITHOUT RESULT IN THE FOREST OF THE NEIGHBORING UDAWADDATEKELE SANCTUARY, CLOSE TO DALADA MALIGAVA, THE TEMPLE HOLDING A GIANT TOOTH SAID TO BE THAT OF BUDDHA.
UNREWARDED BY HIS BLESSED AURA, I TOOK A BUS SOUTH TO THE GEM CENTER OF RATNAPURA, WHERE I CHECKED INTO THE GOVERNMENT REST HOUSE, EAGER TO GET STARTED. SO, I...
...DROPPED MY ARMY-ISSUE DUFFEL BAG IN MY ROOM, SWUNG MY STAINED CANVAS COLLECTOR'S BAG OVER MY SHOULDER
...WALKED DOWN THE STAIRS AND OUT THE BACK ENTRANCE A HUNDRED METERS TO A LINE OF TREES FRINGING THE TOWN RESERVOIR
...LOOKED AROUND, PICKED UP A DEAD TWIG LYING ON THE GROUND, AND BROKE IT OPEN.

...AND STARED AS A COLONY OF SMALL, YELLOW ANTS RAN OUT OVER MY HAND.
ANEURETUS!!
I COULD NOT HAVE BEEN HAPPIER IF I HAD DISCOVERED A PRICELESS RATNAPURA SAP-PHIRE LYING UNCLAIMED ON THE GROUND.
SETTLED BACK IN MY ROOM, I TURNED THE VIAL OF SPECIMENS SLOWLY OVER AND OVER IN MY HAND, LOOKING AT THE FIRST QUEEN, LARVAE, AND SOLDIERS OF THE LIVING *ANEURETINE* EVER SEEN.
"THIS WAS ONE OF THE GREAT THRILLS OF MY LIFE. DINNER TASTED FIT FOR A GOURMET THAT EVENING."

MY DARLING:
IN THE DAYS THAT FOLLOWED, I VENTURED INTO FOREST CLOSER TO ADAM'S PEAK. THOUGH SOMETIMES DELAYED FOR HOURS BY MONSOON DOWNPOURS, THE KIND CALLED GULLY-WASHERS OR FROG-STRANGLERS BACK HOME, I EASILY SECURED MORE COLONIES.
IN A SHORT WHILE, I WAS ABLE TO PUT TOGETHER A PICTURE OF THE SOCIAL LIFE OF THE LAST SURVIVING *ANEURETINE*.
MY FIELD ADVENTURES NOW FINISHED, I CONTINUED ON BY A SECOND ITALIAN LINER (CHEAP FARE) TO GENOA, WHERE I WORKED ON THE ANT COLLECTION OF CARLO EMERY AT THE MUSEO CIVICO DI STORIA NATURALE.
THEN I PROCEEDED BY TRAIN TO SEE THE SIGHTS OF SWITZERLAND AND FRANCE AND FINALLY TO ENGLAND.
GENOA
GENEVA
PARIS
LONDON

ON SEPTEMBER 5, 1955, I FLEW TO NEW YORK, TWENTY POUNDS UNDERWEIGHT, AND TINTED FAINT YELLOW FROM THE ANTIMALARIAL DRUG QUINACRINE.

WE GOT MARRIED IN BOSTON'S ST. CECILIA.

OUR MARRIAGE HAS BEEN HAPPY AND ENDURING.

THE FOLLOWING WINTER, I WAS OFFERED AN ASSISTANT PROFESSORSHIP IN HARVARD'S DEPARTMENT OF BIOLOGY. THE APPOINTMENT WAS ONLY FOR FIVE YEARS—NOT TENURE TRACK.

NO, THE IMPERMANENCE DOESN'T FAZE ME. I HAVE A WORLD-CLASS COLLECTION AND LIBRARY AT MY DISPOSAL, AND TIME ENOUGH TO EXPAND MY RESEARCH.
AND WE'RE YOUNG!

PARTWAY INTO THE FIRST YEAR, THOUGH, MY NERVE BEGAN TO FAIL.
I FEEL... DISPOSABLE.
LIKE ALL ASSISTANT PROFESSORS AT HARVARD, YOU MEAN?
AND OBVIOUSLY, I WAS.

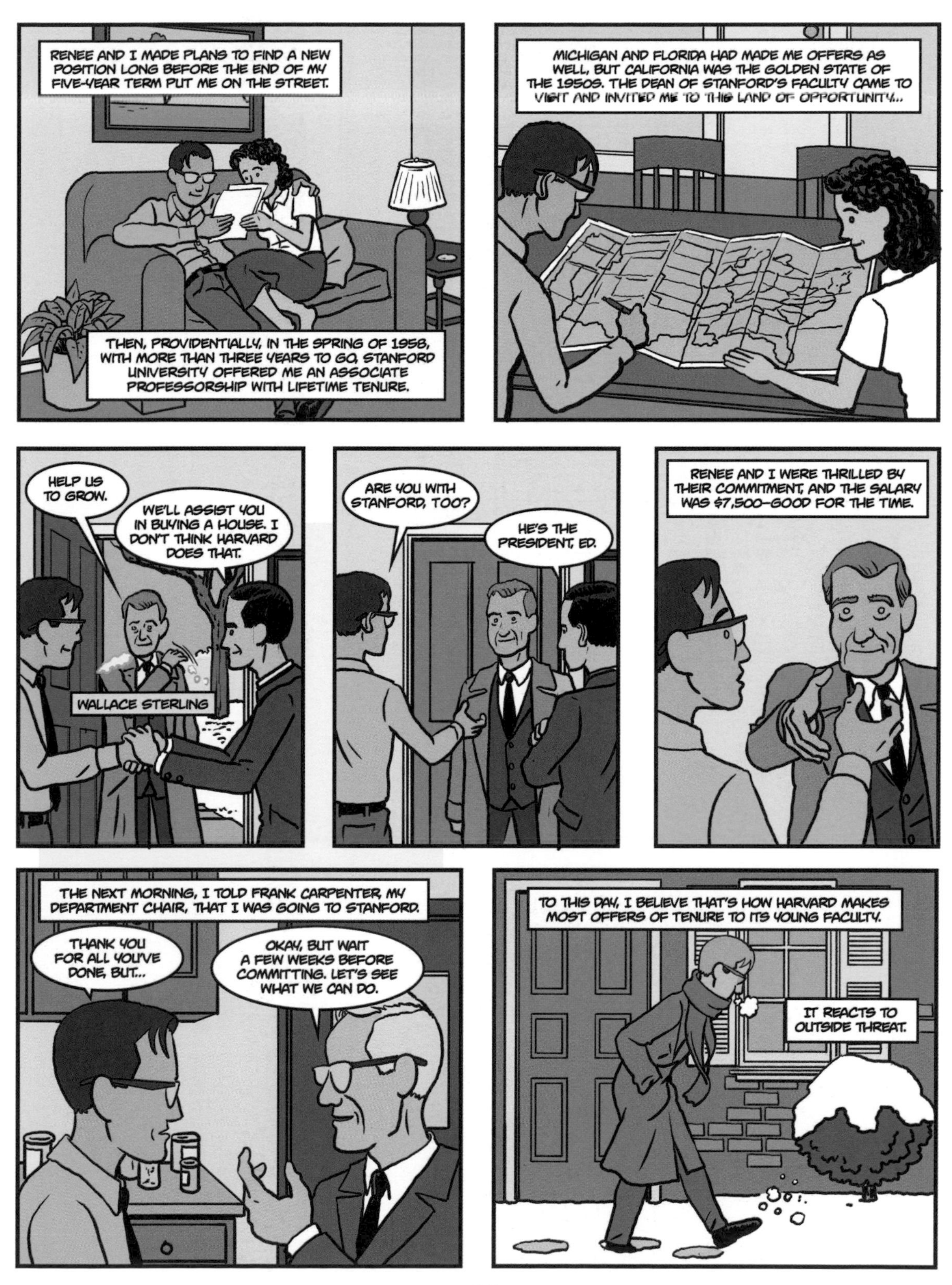
RENEE AND I MADE PLANS TO FIND A NEW POSITION LONG BEFORE THE END OF MY FIVE-YEAR TERM PUT ME ON THE STREET.
THEN, PROVIDENTIALLY, IN THE SPRING OF 1958, WITH MORE THAN THREE YEARS TO GO, STANFORD UNIVERSITY OFFERED ME AN ASSOCIATE PROFESSORSHIP WITH LIFETIME TENURE.
MICHIGAN AND FLORIDA HAD MADE ME OFFERS AS WELL, BUT CALIFORNIA WAS THE GOLDEN STATE OF THE 1950S. THE DEAN OF STANFORD'S FACULTY CAME TO VISIT AND INVITED ME TO THIS LAND OF OPPORTUNITY...
HELP US TO GROW.
WE'LL ASSIST YOU IN BUYING A HOUSE. I DON'T THINK HARVARD DOES THAT.
WALLACE STERLING
ARE YOU WITH STANFORD, TOO?
HE'S THE PRESIDENT, ED.
RENEE AND I WERE THRILLED BY THEIR COMMITMENT, AND THE SALARY WAS $7,500—GOOD FOR THE TIME.
THE NEXT MORNING, I TOLD FRANK CARPENTER, MY DEPARTMENT CHAIR, THAT I WAS GOING TO STANFORD.
THANK YOU FOR ALL YOU'VE DONE, BUT...
OKAY, BUT WAIT A FEW WEEKS BEFORE COMMITTING. LET'S SEE WHAT WE CAN DO.
TO THIS DAY, I BELIEVE THAT'S HOW HARVARD MAKES MOST OFFERS OF TENURE TO ITS YOUNG FACULTY.
IT REACTS TO OUTSIDE THREAT.

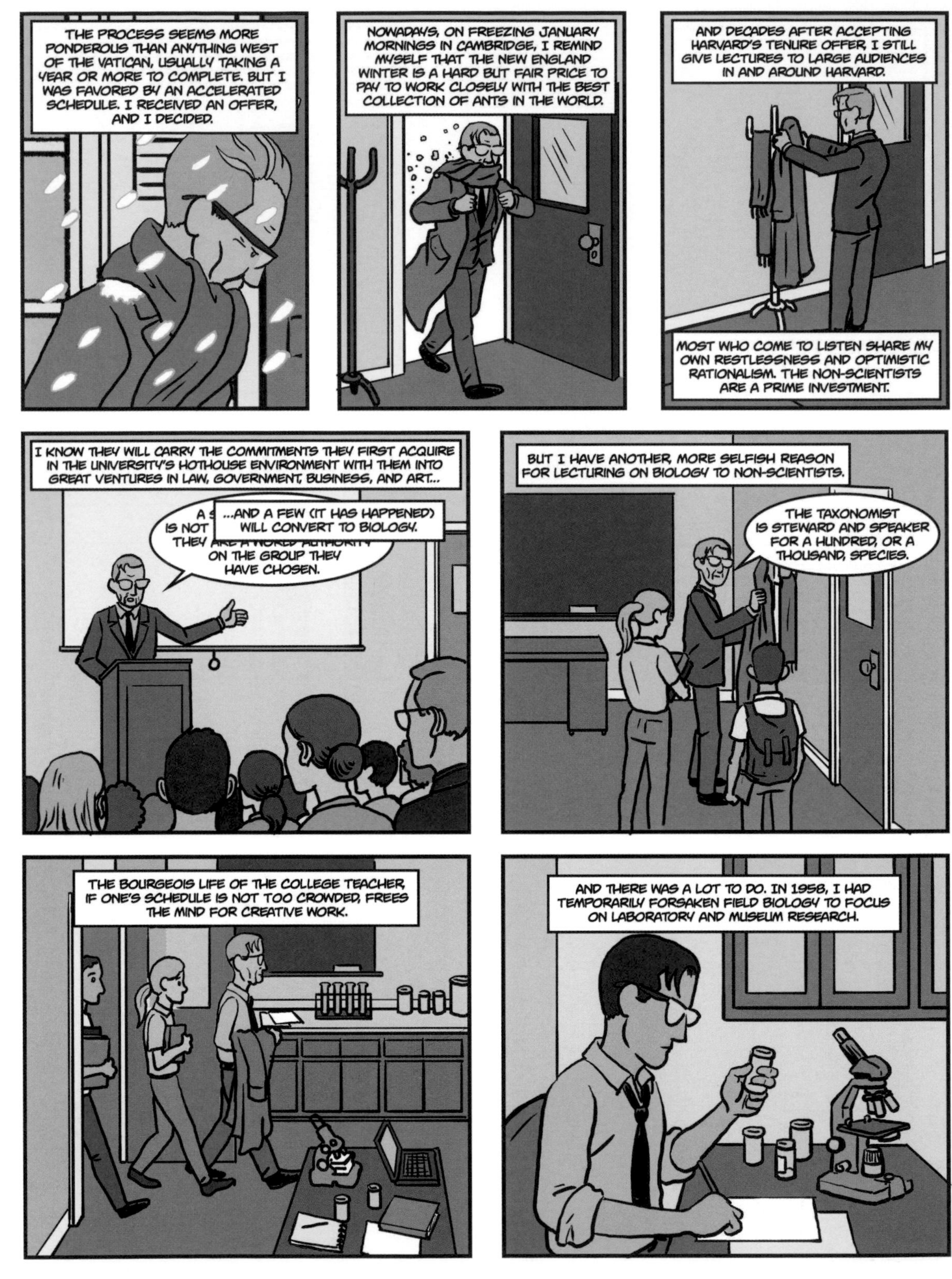
THE PROCESS SEEMS MORE PONDEROUS THAN ANYTHING WEST OF THE VATICAN, USUALLY TAKING A YEAR OR MORE TO COMPLETE. BUT I WAS FAVORED BY AN ACCELERATED SCHEDULE. I RECEIVED AN OFFER, AND I DECIDED.
NOWADAYS, ON FREEZING JANUARY MORNINGS IN CAMBRIDGE, I REMIND MYSELF THAT THE NEW ENGLAND WINTER IS A HARD BUT FAIR PRICE TO PAY TO WORK CLOSELY WITH THE BEST COLLECTION OF ANTS IN THE WORLD.
AND DECADES AFTER ACCEPTING HARVARD'S TENURE OFFER, I STILL GIVE LECTURES TO LARGE AUDIENCES IN AND AROUND HARVARD.
MOST WHO COME TO LISTEN SHARE MY OWN RESTLESSNESS AND OPTIMISTIC RATIONALISM. THE NON-SCIENTISTS ARE A PRIME INVESTMENT.
I KNOW THEY WILL CARRY THE COMMITMENTS THEY FIRST ACQUIRE IN THE UNIVERSITY'S HOTHOUSE ENVIRONMENT WITH THEM INTO GREAT VENTURES IN LAW, GOVERNMENT, BUSINESS, AND ART...
A S
IS NOT
ON THE GROUP THEY HAVE CHOSEN.
...AND A FEW (IT HAS HAPPENED) WILL CONVERT TO BIOLOGY.
BUT I HAVE ANOTHER, MORE SELFISH REASON FOR LECTURING ON BIOLOGY TO NON-SCIENTISTS.
THE TAXONOMIST IS STEWARD AND SPEAKER FOR A HUNDRED, OR A THOUSAND, SPECIES.
THE BOURGEOIS LIFE OF THE COLLEGE TEACHER, IF ONE'S SCHEDULE IS NOT TOO CROWDED, FREES THE MIND FOR CREATIVE WORK.
AND THERE WAS A LOT TO DO. IN 1958, I HAD TEMPORARILY FORSAKEN FIELD BIOLOGY TO FOCUS ON LABORATORY AND MUSEUM RESEARCH.

MY CENTRAL AIM WAS THE CLASSIFICATION AND ANALYSIS OF THE ANTS OF NEW GUINEA AND THE SURROUNDING REGIONS OF TROPICAL ASIA, AUSTRALIA, AND THE SOUTH PACIFIC.
I EMBARKED ON A BREAD-AND-BUTTER TASK... TIME-CONSUMING, TEDIOUS, FACT-CENTERED.
AND, FOR THIS COMBINATION OF REASONS, VIRTUOUS IN MY OWN MIND.
DURING MY ACCUMULATION OF FACTS ABOUT ANT BIOLOGY, VAPOROUS NOTIONS, CONSTRUCTS, DEFINITIONS, INCHOATE PATTERNS...
...THE PERFECT PHRASE ESCAPES ME...
...BUT THE PATTERNS DRIFTED IN AND OUT OF MY MIND LIKE CELTIC FOG.
MY DAYDREAMS WERE MOSTLY ABOUT THE ORIGINS OF BIOLOGICAL DIVERSITY. MOST TOOK COHERENT FORM ONLY TO PROVE MARGINAL OR UNATTRACTIVE.
A FEW WENT ON TO GAIN ROBUST LIFE IN THE COURSE OF MY DAILY REVERIE. THEY THEN TURNED INTO NARRATIVES, WHICH I BEGAN TO REPEAT TO MYSELF LIKE STORIES.
SUBSPECIES. REAL OR IMAGINED?

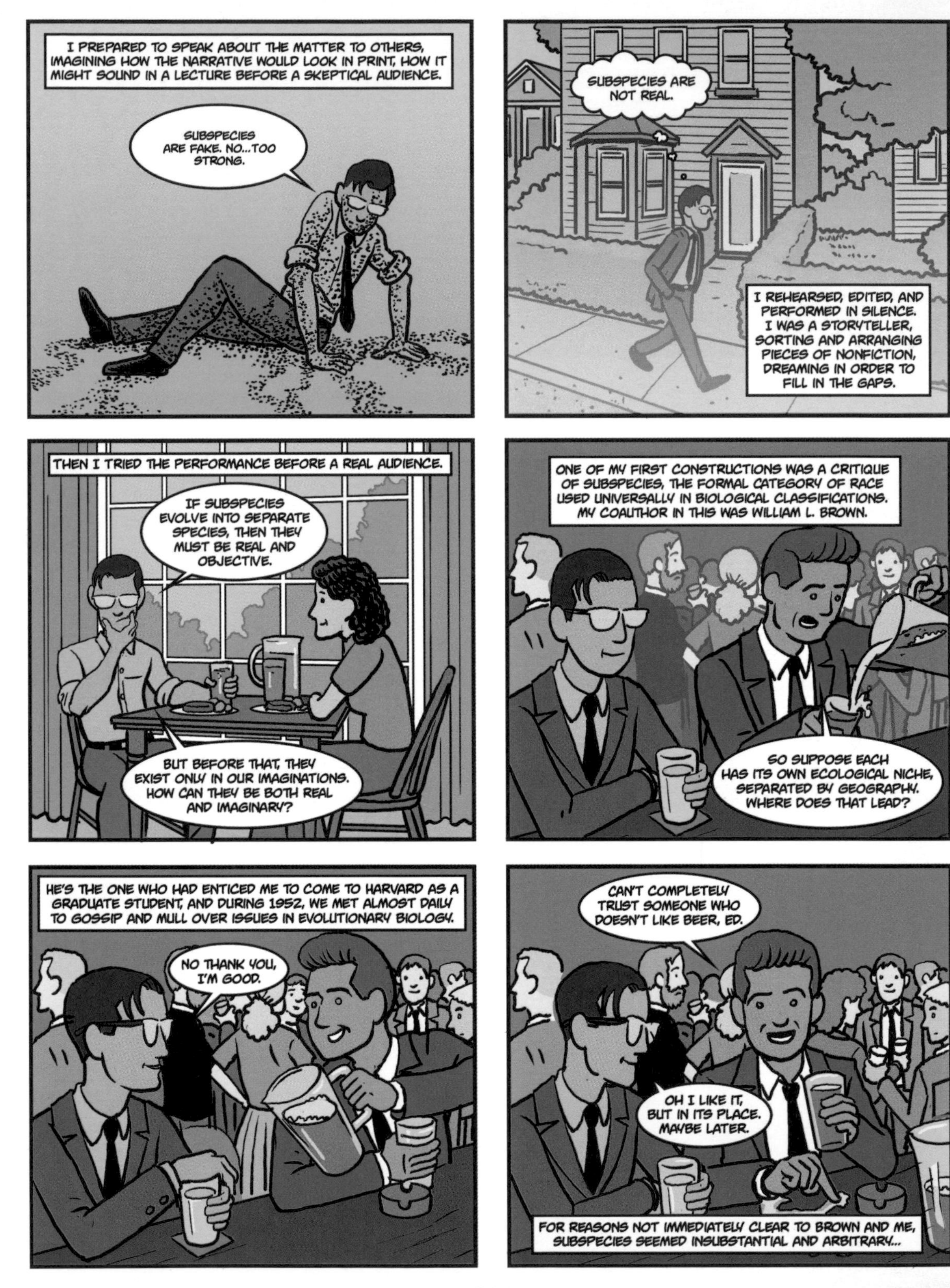
I PREPARED TO SPEAK ABOUT THE MATTER TO OTHERS, IMAGINING HOW THE NARRATIVE WOULD LOOK IN PRINT, HOW IT MIGHT SOUND IN A LECTURE BEFORE A SKEPTICAL AUDIENCE.
SUBSPECIES ARE FAKE. NO...TOO STRONG.
SUBSPECIES ARE NOT REAL.
I REHEARSED, EDITED, AND PERFORMED IN SILENCE. I WAS A STORYTELLER, SORTING AND ARRANGING PIECES OF NONFICTION, DREAMING IN ORDER TO FILL IN THE GAPS.
THEN I TRIED THE PERFORMANCE BEFORE A REAL AUDIENCE.
IF SUBSPECIES EVOLVE INTO SEPARATE SPECIES, THEN THEY MUST BE REAL AND OBJECTIVE.
BUT BEFORE THAT, THEY EXIST ONLY IN OUR IMAGINATIONS. HOW CAN THEY BE BOTH REAL AND IMAGINARY?
ONE OF MY FIRST CONSTRUCTIONS WAS A CRITIQUE OF SUBSPECIES, THE FORMAL CATEGORY OF RACE USED UNIVERSALLY IN BIOLOGICAL CLASSIFICATIONS. MY COAUTHOR IN THIS WAS WILLIAM L. BROWN.
SO SUPPOSE EACH HAS ITS OWN ECOLOGICAL NICHE, SEPARATED BY GEOGRAPHY. WHERE DOES THAT LEAD?
HE'S THE ONE WHO HAD ENTICED ME TO COME TO HARVARD AS A GRADUATE STUDENT, AND DURING 1952, WE MET ALMOST DAILY TO GOSSIP AND MULL OVER ISSUES IN EVOLUTIONARY BIOLOGY.
NO THANK YOU, I'M GOOD.
CAN'T COMPLETELY TRUST SOMEONE WHO DOESN'T LIKE BEER, ED.
OH I LIKE IT, BUT IN ITS PLACE. MAYBE LATER.
FOR REASONS NOT IMMEDIATELY CLEAR TO BROWN AND ME, SUBSPECIES SEEMED INSUBSTANTIAL AND ARBITRARY...

THE DISCORDANCE CAN BE MOST IMMEDIATELY UNDERSTOOD WITH AN IMAGINARY BUT TYPICAL EXAMPLE...
IN NATURE COLOR MIGHT VARY EAST TO WEST AND SIZE FROM NORTH TO SOUTH.
AND AN EXTRA BAND APPEARS ON THE HIND WING IN A FEW LOCALITIES NEAR THE CENTER.
AND SO ON FOR ANY NUMBER OF TRAITS A TAXONOMIST MIGHT CHOOSE FROM AN ALMOST ENDLESS LIST AVAILABLE FOR CLASSIFICATION.
PICK COLOR, AND YOU HAVE TWO EAST-WEST RACES.
PICK COLOR PLUS SIZE, AND FOUR RACES IN A QUADRANT COME INTO EXISTENCE.
ADD A HIND-WING BAND, AND THE NUMBER OF RACES CAN DOUBLE AGAIN. HENCE, SUBSPECIES ARE ARBITRARY.

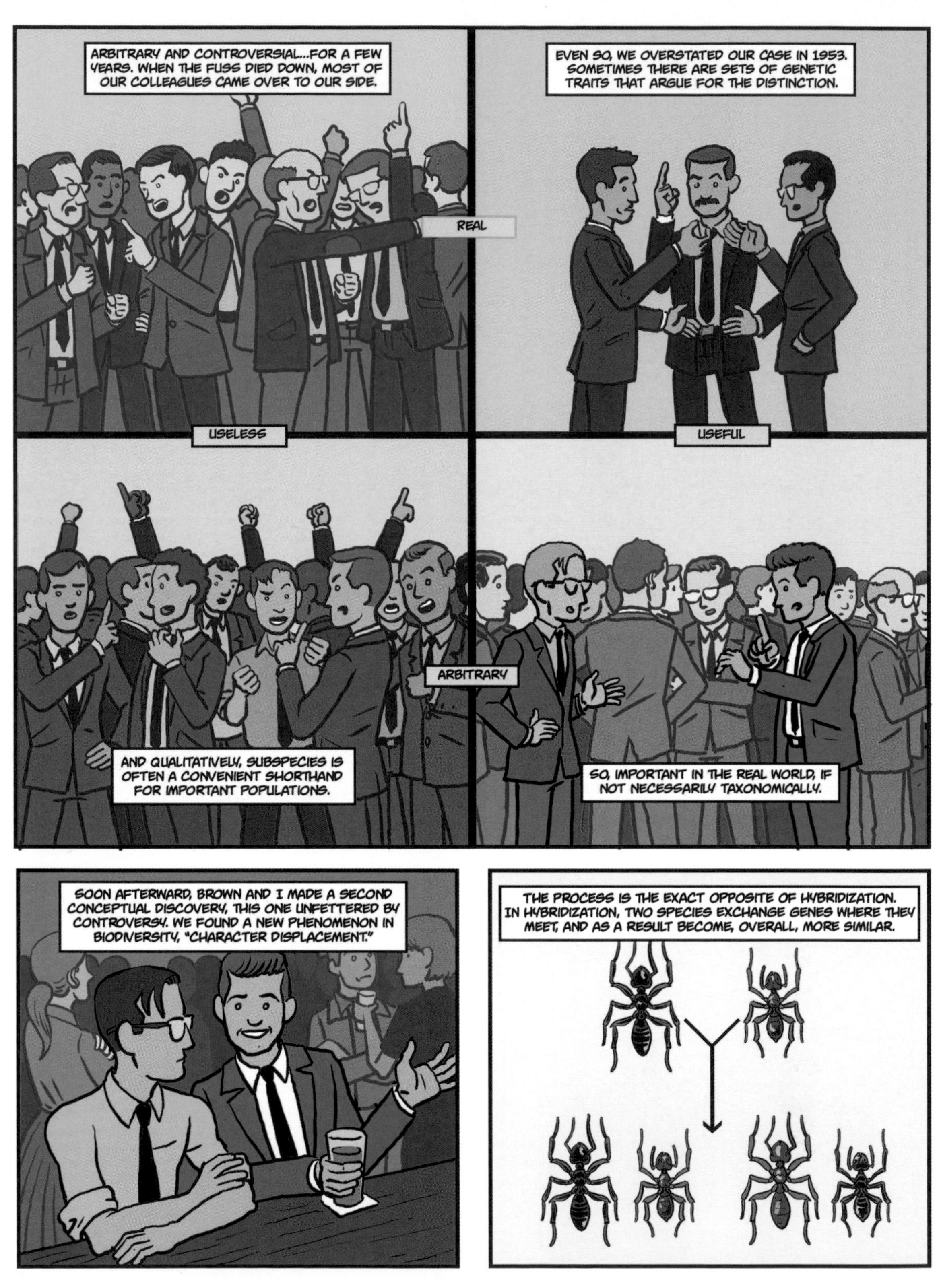
ARBITRARY AND CONTROVERSIAL...FOR A FEW YEARS. WHEN THE FUSS DIED DOWN, MOST OF OUR COLLEAGUES CAME OVER TO OUR SIDE.
REAL
EVEN SO, WE OVERSTATED OUR CASE IN 1953. SOMETIMES THERE ARE SETS OF GENETIC TRAITS THAT ARGUE FOR THE DISTINCTION.
USELESS
USEFUL
ARBITRARY
AND QUALITATIVELY, SUBSPECIES IS OFTEN A CONVENIENT SHORTHAND FOR IMPORTANT POPULATIONS.
SO, IMPORTANT IN THE REAL WORLD, IF NOT NECESSARILY TAXONOMICALLY.
SOON AFTERWARD, BROWN AND I MADE A SECOND CONCEPTUAL DISCOVERY, THIS ONE UNFETTERED BY CONTROVERSY. WE FOUND A NEW PHENOMENON IN BIODIVERSITY, "CHARACTER DISPLACEMENT."
THE PROCESS IS THE EXACT OPPOSITE OF HYBRIDIZATION. IN HYBRIDIZATION, TWO SPECIES EXCHANGE GENES WHERE THEY MEET, AND AS A RESULT BECOME, OVERALL, MORE SIMILAR.

IN CHARACTER DISPLACEMENT TWO SPECIES SPRING APART WHERE THEY MEET, LIKE PARTICLES WITH THE SAME CHARGE.
THE BRITISH ORNITHOLOGIST DAVID LACK HAD ALREADY DELINEATED THIS IN HIS 1947 STUDY OF DARWIN'S FINCHES ON THE GALAPAGOS ISLANDS...
BROWN AND I SHOWED THAT THE REPELLANT EFFECT IS WIDESPREAD AND IS CAUSED, ACCORDING TO THE SPECIES PAIR CONSIDERED, BY EITHER COMPETITION OR THE ACTIVE AVOIDANCE OF HYBRIDIZATION.
CHARACTER DISPLACEMENT, WE ALSO REALIZED, IS ONE MEANS BY WHICH SPECIES CAN BE PACKED TOGETHER MORE TIGHTLY IN ECOSYSTEMS.
THE EVOLUTION OF GREATER DIFFERENCES BETWEEN SPECIES REDUCES THE CHANCE THAT ONE OF THEM WILL ERASE THE OTHER THROUGH COMPETITION OR HYBRIDIZATION.
THE BETTER THE MUTUAL ADJUSTMENT THAT AVOIDS COMPETITION AND HYBRIDIZATION, THE MORE SPECIES THAT CAN LIVE TOGETHER INDEFINITELY.
HENCE, THE BIODIVERSITY WILL BE RICHER AS AN OUTCOME OF EVOLUTION IN THE COMMUNITY AS A WHOLE.

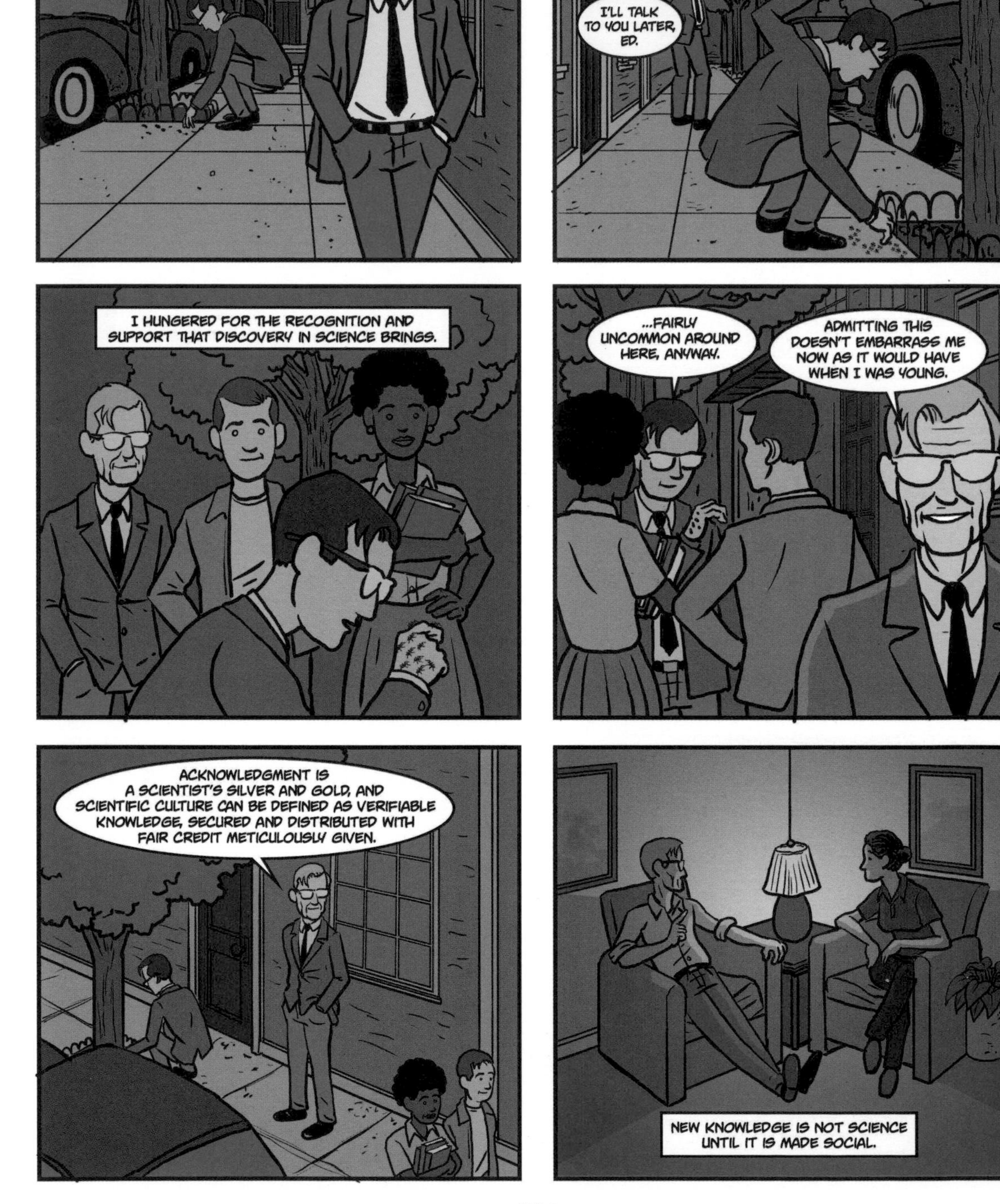
BY THE LATE 1950S, I HAD CLEARLY GROWN MORE INTERESTED IN THEORY, BUT THE BOY INSIDE STILL MADE MY CAREER DECISIONS.
I JUST WANTED TO BE THE FIRST TO FIND *SOMETHING*, ANYTHING, THE MORE IMPORTANT THE BETTER, BUT SOMETHING AS OFTEN AS POSSIBLE. AND TO OWN IT A LITTLE WHILE BEFORE RELINQUISHING IT TO OTHERS.
I'LL TALK TO YOU LATER, ED.
I HUNGERED FOR THE RECOGNITION AND SUPPORT THAT DISCOVERY IN SCIENCE BRINGS.
...FAIRLY UNCOMMON AROUND HERE, ANYWAY.
ADMITTING THIS DOESN'T EMBARRASS ME NOW AS IT WOULD HAVE WHEN I WAS YOUNG.
ACKNOWLEDGMENT IS A SCIENTIST'S SILVER AND GOLD, AND SCIENTIFIC CULTURE CAN BE DEFINED AS VERIFIABLE KNOWLEDGE, SECURED AND DISTRIBUTED WITH FAIR CREDIT METICULOUSLY GIVEN.
NEW KNOWLEDGE IS NOT SCIENCE UNTIL IT IS MADE SOCIAL.

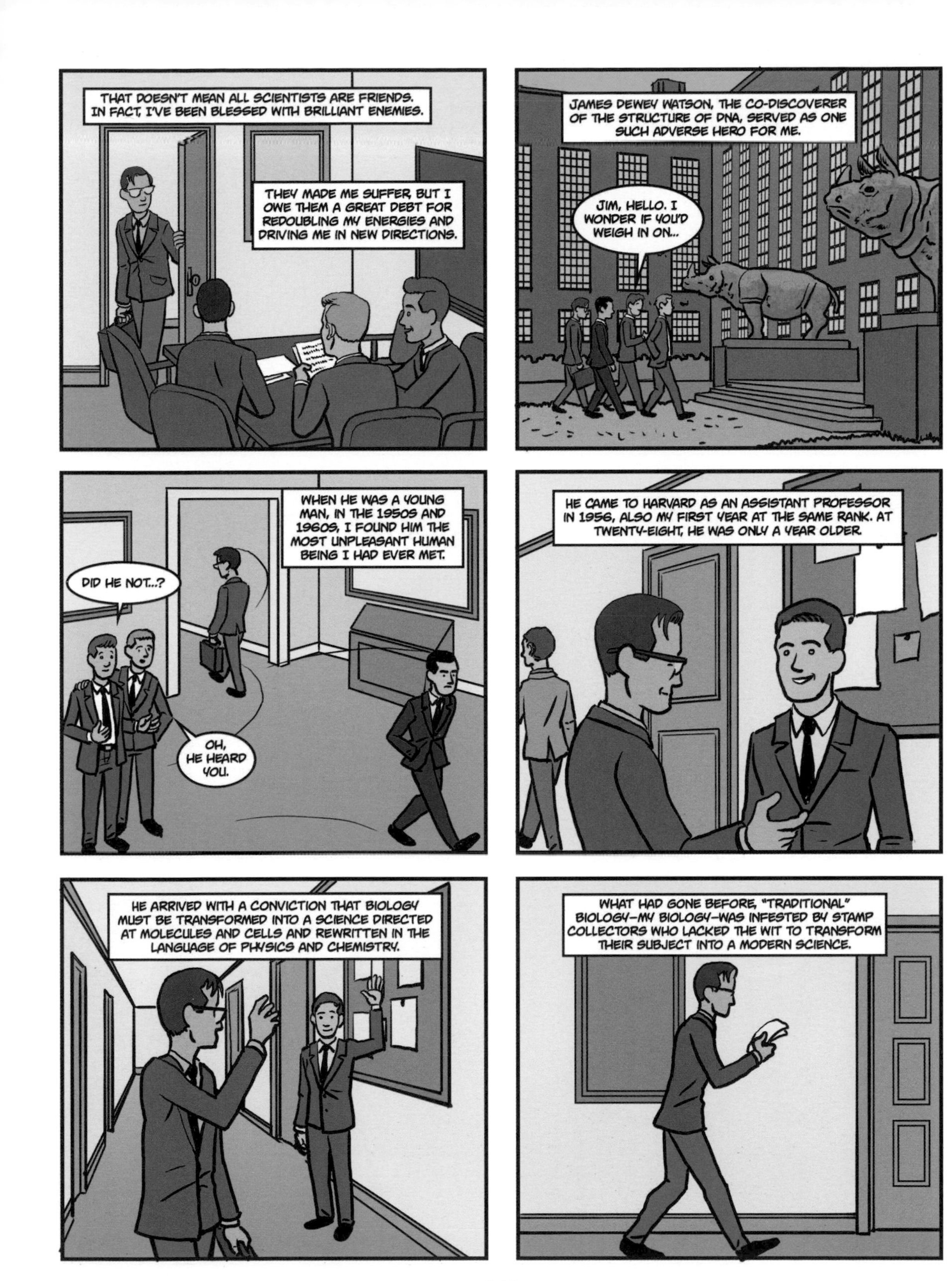
THAT DOESN'T MEAN ALL SCIENTISTS ARE FRIENDS. IN FACT, I'VE BEEN BLESSED WITH BRILLIANT ENEMIES.
THEY MADE ME SUFFER, BUT I OWE THEM A GREAT DEBT FOR REDOUBLING MY ENERGIES AND DRIVING ME IN NEW DIRECTIONS.
JAMES DEWEY WATSON, THE CO-DISCOVERER OF THE STRUCTURE OF DNA, SERVED AS ONE SUCH ADVERSE HERO FOR ME.
JIM, HELLO. I WONDER IF YOU'D WEIGH IN ON...
WHEN HE WAS A YOUNG MAN, IN THE 1950S AND 1960S, I FOUND HIM THE MOST UNPLEASANT HUMAN BEING I HAD EVER MET.
DID HE NOT...?
OH, HE HEARD YOU.
HE CAME TO HARVARD AS AN ASSISTANT PROFESSOR IN 1956, ALSO MY FIRST YEAR AT THE SAME RANK. AT TWENTY-EIGHT, HE WAS ONLY A YEAR OLDER.
HE ARRIVED WITH A CONVICTION THAT BIOLOGY MUST BE TRANSFORMED INTO A SCIENCE DIRECTED AT MOLECULES AND CELLS AND REWRITTEN IN THE LANGUAGE OF PHYSICS AND CHEMISTRY.
WHAT HAD GONE BEFORE, "TRADITIONAL" BIOLOGY—MY BIOLOGY—WAS INFESTED BY STAMP COLLECTORS WHO LACKED THE WIT TO TRANSFORM THEIR SUBJECT INTO A MODERN SCIENCE.

WATSON TREATED MOST OF THE OTHER TWENTY-FOUR MEMBERS OF THE DEPARTMENT OF BIOLOGY WITH A REVOLUTIONARY'S FERVENT DISRESPECT.

AT DEPARTMENT MEETINGS, WATSON RADIATED CONTEMPT IN ALL DIRECTIONS...TOLERATED BECAUSE OF THE GREATNESS OF THE DISCOVERY HE HAD MADE, AND BECAUSE OF ITS GATHERING AFTERMATH.
I WOULD LIKE TO PROPOSE A JOINT APPOINTMENT.

IN HIS OWN MIND, APPARENTLY, HE WAS HONEST JIM, AS HE ORIGINALLY TITLED HIS MEMOIR OF THE DISCOVERY.
MY STORY
DISCOVERY
THE DOUBLE HELIX

WATSON'S ATTITUDE WAS PARTICULARLY PAINFUL FOR ME.
THE DEPARTMENT NEEDS MORE ENVIRONMENTAL BIOLOGISTS, FOR BALANCE.
WE SHOULD AT LEAST DOUBLE THE NUMBER FROM ONE...

...ME...
...TO TWO.

I HAVE A CANDIDATE IN MIND, AN ECOLOGIST WHO'S ALREADY ON CAMPUS IN THE GRADUATE SCHOOL OF DESIGN. WE COULD OFFER HIM JOINT MEMBERSHIP HERE.
ARE THEY OUT OF THEIR MINDS?

WHAT DO YOU MEAN?
!
ANYONE WHO WOULD HIRE AN ECOLOGIST IS OUT OF HIS MIND.
THUS SPOKE THE AVATAR OF MOLECULAR BIOLOGY.
UMM. RIGHT. THANKS ED. WE'LL TAKE THIS UP LATER.
NO ONE ECHOED WATSON, BUT NO ONE DEFENDED THE NOMINATION EITHER. THE IDEA WAS SET ASIDE, AND NOT SPOKEN OF AGAIN. THE MOLECULAR BIOLOGISTS SPLIT OFF TO FORM A DEPARTMENT OF THEIR OWN.
ED, HEY.
I DON'T THINK WE SHOULD SAY "ECOLOGY" ANYMORE. IT'S BECOME A DIRTY WORD.
REALLY?
SURE ENOUGH, FOR MOST OF THE FOLLOWING DECADE WE LARGELY STOPPED USING THE WORD "ECOLOGY."
REALLY.
ONLY LATER DID I SENSE THE ANTHROPOLOGICAL SIGNIFICANCE. WHEN ONE CULTURE SETS OUT TO ERASE ANOTHER, THE FIRST THING ITS RULERS BANISH IS USE OF THE NATIVE TONGUE.
THE MOLECULAR WARS WERE ON, AND BY THE LATE 1950S, THE ATMOSPHERE HAD BECOME TOO STIFLING FOR US TO PLAN THE FUTURE OF HARVARD BIOLOGY IN ORDINARY MEETINGS.

WHEN TRIBES REACH A CERTAIN SIZE AND DENSITY, THEY SPLIT, AND ONE GROUP EMIGRATES TO A NEW TERRITORY. SO THE PROFESSORS IN ORGANISMIC AND EVOLUTIONARY BIOLOGY PREPARED TO EXIT.
AMONG THE YAMOMAMI OF BRAZIL AND VENEZUELA, THE MOMENT OF FISSION CAN BE JUDGED TO BE CLOSE AT HAND WHEN THERE IS A SHARP INCREASE IN AX FIGHTING.
BY THE FALL OF 1962, WE HAD SPLIT OFF A COMMITTEE ON EVOLUTIONARY BIOLOGY. BUT THERE WERE STILL CASUALTIES. THESE DAYS SYSTEMATISTS, THE SOLITARY EXPERTS ON GROUPS OF ORGANISMS, HAVE UNFORTUNATELY BEEN LARGELY ELIMINATED.
CEBINAE
LYNX RUFUS
EUTAMIAS SPP
THAT IS THE WORST SINGLE DAMAGE CAUSED BY THE MOLECULAR REVOLUTION.
ECOLOGISTS, PUSHED TO THE MARGIN FOR YEARS, HAVE BEGUN A RESURGENCE THROUGH THE WIDESPREAD RECOGNITION OF THE GLOBAL ENVIRONMENTAL CRISIS.
MOLECULAR BIOLOGISTS, AS THEY PROMISED, HAVE TAKEN UP EVOLUTIONARY STUDIES, MAKING IMPORTANT CONTRIBUTIONS WHENEVER THEY CAN FIND SYSTEMATISTS TO TELL THEM THE NAMES OF ORGANISMS.
THE SURVIVING EVOLUTIONARY BIOLOGISTS ROUTINELY USE MOLECULAR DATA TO PURSUE THEIR DARWINIAN AGENDA.

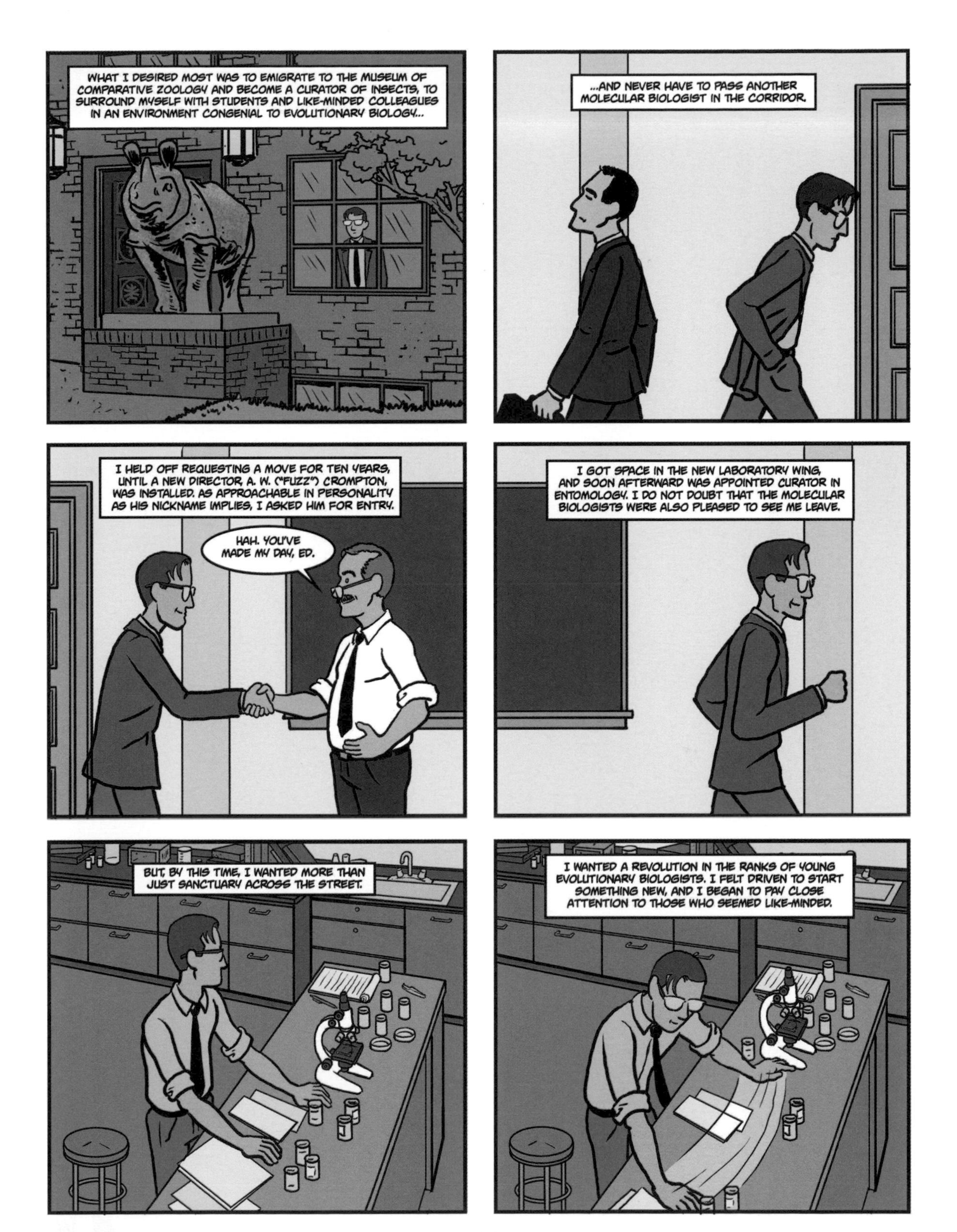
WHAT I DESIRED MOST WAS TO EMIGRATE TO THE MUSEUM OF COMPARATIVE ZOOLOGY AND BECOME A CURATOR OF INSECTS, TO SURROUND MYSELF WITH STUDENTS AND LIKE-MINDED COLLEAGUES IN AN ENVIRONMENT CONGENIAL TO EVOLUTIONARY BIOLOGY...
...AND NEVER HAVE TO PASS ANOTHER MOLECULAR BIOLOGIST IN THE CORRIDOR.
I HELD OFF REQUESTING A MOVE FOR TEN YEARS, UNTIL A NEW DIRECTOR, A. W. ("FUZZ") CROMPTON, WAS INSTALLED. AS APPROACHABLE IN PERSONALITY AS HIS NICKNAME IMPLIES, I ASKED HIM FOR ENTRY.
HAH. YOU'VE MADE MY DAY, ED.
I GOT SPACE IN THE NEW LABORATORY WING, AND SOON AFTERWARD WAS APPOINTED CURATOR IN ENTOMOLOGY. I DO NOT DOUBT THAT THE MOLECULAR BIOLOGISTS WERE ALSO PLEASED TO SEE ME LEAVE.
BUT, BY THIS TIME, I WANTED MORE THAN JUST SANCTUARY ACROSS THE STREET.
I WANTED A REVOLUTION IN THE RANKS OF YOUNG EVOLUTIONARY BIOLOGISTS. I FELT DRIVEN TO START SOMETHING NEW, AND I BEGAN TO PAY CLOSE ATTENTION TO THOSE WHO SEEMED LIKE-MINDED.

AMONG THEM WAS LARRY SLOBODKIN, THEN AN ASSISTANT PROFESSOR AT THE UNIVERSITY OF MICHIGAN AND A RISING STAR IN THE ADMITTEDLY STILL DEPAUPERATE FIELD OF AMERICAN ECOLOGY.
HIS REPUTATION AS A RESEARCHER HAS BEEN SECURELY BASED ON A SERIES OF ECLECTIC STUDIES—THE RED TIDE PHENOMENON, MEASUREMENT OF ENERGY TRANSFER IN ECOSYSTEMS BY THE USE OF THE BOMB CALORIMETER...
KARENIA BREVIS
...AND IN THE REALM OF THEORY, HE ELABORATED THE CONCEPT OF A BALANCING RELATIONSHIP BETWEEN THE "PRUDENT PREDATOR" AND "EFFICIENT PREY."
SLOBODKIN WAS HEAVILY INFLUENCED BY HIS PH.D. ADVISER, G. EVELYN HUTCHINSON, HIMSELF AS DIFFERENT FROM SLOBODKIN AND ME AS LARRY AND I WERE FROM EACH OTHER.
HE WAS A FREE SPIRIT, AN ECLECTICIST WHO PROVED BRILLIANT AT FITTING PIECES TOGETHER INTO LARGE CONCEPTS AND WHO NEVER SEEMED TO HAVE MET A FACT HE DIDN'T LIKE OR COULDN'T USE.
AH, SO IT'S NOT SO MUCH THAT THE PREY ARE EFFICIENT, BUT THAT THE PRUDENT PREDATOR DOESN'T OVER-EXPLOIT THEM EVEN IF IT'S IN THEIR SHORTTERM ADVANTAGE TO DO SO.
IT'S ALMOST LIKE THEY WERE PLANNING FOR THE FUTURE. OR HAD A SOCIAL CONSCIENCE. OR BOTH.
HUTCHINSON'S INSIGHTS WERE DEEP AND ORIGINAL, AND HE DESERVES TO BE CALLED THE FATHER OF EVOLUTIONARY ECOLOGY.

LIKE MOST SUCCESSFUL IDEAS IN SCIENCE, HIS NOTION OF WHAT CAME TO BE CALLED A "HUTCHINSON NICHE" IS A SIMPLE ONE.
SO, HERE'S HOW ONE MIGHT CONCEPTUALIZE IT...
AND SO ON DOWN A LIST AS LONG AS THE BIOLOGIST WISHES TO MAKE IT. WE VIEW THE SPECIES AS LIVING WITHIN A SPACE DEFINED BY THE LIMITS OF THESE BIOLOGICAL QUALITIES, EACH PLACED IN TURN ON A SEPARATE SCALE.
THROUGH THEM, I CAME TO APPRECIATE HOW ENVIRONMENTAL SCIENCE MIGHT BE BETTER MESHED WITH BIOGEOGRAPHY AND THE STUDY OF EVOLUTION, AND I GAINED MORE CONFIDENCE IN THE INTELLECTUAL INDEPENDENCE OF EVOLUTIONARY BIOLOGY.

A SPECIES CAN BE USEFULLY
OF TEMPERATURES IN WHICH IT IS ABLE
REPRODUCE, THE RANGE OF PREY ITEMS IT
THE SEASON IN WHICH IT IS ACTIVE, THE
THE DAY DURING WHICH IT FEEDS

SO, IN SHORT, YOU'RE SAYING THE SPECIES' NICHE IS AN N-DIMENSIONAL HYPERSPACE.
YES. MAYBE?
HUTCHINSON AND SLOBODKIN WERE WHAT TODAY ARE CALLED EVOLUTIONARY ECOLOGISTS. IN MY FORMATIVE YEARS, THEY CAUSED ME TO TRY TO BECOME ONE AS WELL.

I WAS ENCOURAGED TO DRAW CLOSER TO THE CENTRAL PROBLEM OF THE BALANCE OF SPECIES, WHICH WAS TO BE MY MAIN PREOCCUPATION DURING THE 1960S, AS THE MOLECULAR WARS SUBSIDED TO THEIR AMBIGUOUS CONCLUSION.

AT AROUND THE TIME ALL THIS WAS HAPPENING, I TOO WAS DRAWN, AS A RESULT OF MY EXPERIENCE IN ZOOGEOGRAPHY, TO THE IDEA OF THE FOUNTAINHEADS OF EVOLUTION.
SERIOUS DISCUSSION OF WHERE DOMINANT GROUPS RISE AND HOW THEY SPREAD TO THE REST OF THE WORLD HAD BEGUN IN 1915 WITH WILLIAM DILLER MATTHEW, CURATOR IN INVERTEBRATE PALEONTOLOGY AT THE AMERICAN MUSEUM OF NATURAL HISTORY.
IN HIS BOOK *CLIMATE AND EVOLUTION*, HE CONSTRUCTED A PICTURE OF DOMINANCE IN MAMMALS AND OTHER VERTEBRATES. MATTHEW INSTRUCTED THE READER TO LOOK AT A NORTH POLAR PROJECTION OF THE GLOBE.
EUROPE, ASIA, AND NORTH AMERICA ARE SO CLOSE TOGETHER AS TO FORM A SINGLE SUPERCONTINENT.
HE CITED EVIDENCE FROM HIS OWN AND PREVIOUS FOSSIL STUDIES, AND AUDACIOUSLY POSITED THAT THE DOMINANT GROUPS AROSE ON THIS SUPERCONTINENT AND SPREAD OUTWARD.
THEY DISPLACED FORMERLY DOMINANT GROUPS SOUTHWARD TO THE PERIPHERAL REGIONS OF TROPICAL ASIA, AFRICA, AND SOUTH AMERICA.
IN OUR OWN TIME, THE WINNERS INCLUDE DEER, CAMELS, PIGS, AND THE MOST FAMILIAR OF THE RATS AND MICE.
"AMONG THE LOSERS RETREATING TO THE EDGES OF THE WORLD ARE HORSES, TAPIRS, AND RHINOCEROSES."

THEN ALONG CAME PHILIP DARLINGTON WITH A NEW TWIST, PRESENTED FIRST IN A 1948 ARTICLE AND LATER IN HIS 1957 FULL-DRESS TREATISE *ZOOGEOGRAPHY: THE GEOGRAPHICAL DISTRIBUTION OF ANIMALS.*
MATTHEW WAS ONLY HALF RIGHT. THE FOSSIL REMAINS HE STUDIED WERE BIASED TOWARD THE NORTHERN HEMISPHERE.
THAT WAS NATURAL, SINCE MOST COLLECTING HAD BEEN CONCENTRATED THERE DURING THE EARLY YEARS OF PALEONTOLOGY.
THIRTY YEARS AFTER HIS BOOK *CLIMATE AND EVOLUTION*, THOUGH, WE HAVE MORE DATA ON FOSSILS TO EXAMINE, AND THESE COME FROM ALL OVER THE WORLD, INCLUDING MATTHEW'S "PERIPHERAL" AREAS.
WE ALSO MUST EXAMINE MORE CAREFULLY THE DISTRIBUTION OF LIVING GROUPS, ESPECIALLY FISHES, FROGS, AND OTHER COLD-BLOODED VERTEBRATES, ANIMALS FOR WHICH MOST OF THE NEW EVIDENCE IS AVAILABLE.
WHEN WE PUT ALL THE PIECES TOGETHER, WE SEE THAT THE EVOLUTIONARY CRUCIBLE IS NOT THE NORTH TEMPERATE LAND MASS, BUT THE OLD WORLD TROPICS.
FOR THE PAST 50 MILLION YEARS OR SO, GROUPS OF VERTEBRATES HAVE ARISEN IN THE VAST REGION THAT COMPRISES SOUTHERN ASIA, SUB-SAHARAN AFRICA, AND, UNTIL RECENT GEOLOGICAL TIMES, MUCH OF THE MIDDLE EAST.
THE MOST DOMINANT ANIMAL GROUPS PRESSED NORTHWARD TO EUROPE AND SIBERIA, ACROSS THE BERING SEA, A BARRIER PERIODICALLY BREACHED BY THE RISE OF ISTHMUSES, AND INTO THE NEW WORLD.

PEOPLE LIVING IN NORTH AMERICA OR EUROPE TODAY NEED ONLY LOOK AROUND TO SEE THE CURRENT HEGEMONIC GROUPS.
THEIR SPECIES ARE EXTENDING THEIR RANGES INTO, NOT OUT OF, THE HARSHER CLIMATES.
THIS PROBLEM OF THE BIOLOGICAL CAUSE OF DOMINANCE WAS NOT CLEARLY IN MY MIND AS I BEGAN MY OWN BIOGEOGRAPHIC STUDY ON THE ANTS OF NEW GUINEA AND SURROUNDING REGIONS.
BUT MATTHEW AND DARLINGTON HAD PRIMED ME TO FORMULATE IT.
ALL I NEEDED, I REALIZE TODAY, WAS A SMALL SET OF DATA TO FALL INTO PLACE IN ORDER FOR THE QUESTION TO FORM SOMEWHERE IN MY SUBCONSCIOUS.
I HAD COLLECTED A *LOT* OF DATA ON THE PLACES THE ANTS OF THE WESTERN PACIFIC LIVED, THEIR NEST SITES, THEIR COLONY SIZE, WHAT THEY ATE...

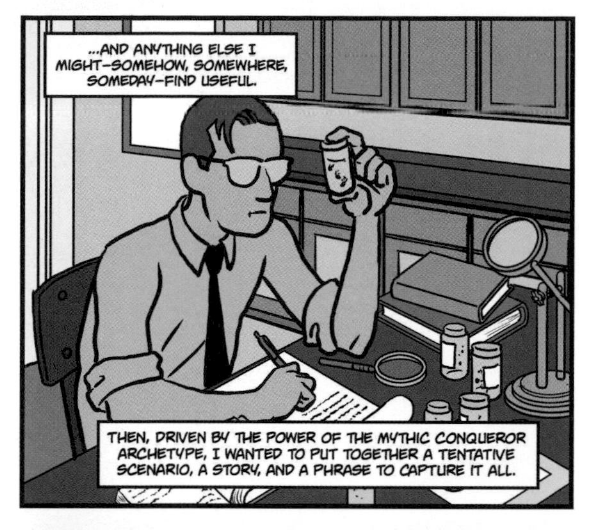
...AND ANYTHING ELSE I MIGHT—SOMEHOW, SOMEWHERE, SOMEDAY—FIND USEFUL.
THEN, DRIVEN BY THE POWER OF THE MYTHIC CONQUEROR ARCHETYPE, I WANTED TO PUT TOGETHER A TENTATIVE SCENARIO, A STORY, AND A PHRASE TO CAPTURE IT ALL.

"AND AS IMAGINATION BODIES FORTH THE FORMS OF THINGS UNKNOWN, THE POET'S PEN TURNS THEM INTO SHAPES, AND GIVES TO AIRY NOTHING A LOCAL HABITATION AND A NAME."

AND A PATTERN DID EMERGE. EVOLUTIONARY BIOLOGY ALWAYS YIELDS PATTERNS IF YOU LOOK HARD ENOUGH, BECAUSE THERE ARE A HUNDRED PARAMETERS AND A THOUSAND PATTERNS AWAITING EXAMINATION.

IT BECAME CLEAR AS I MAPPED RANGES OF ONE SPECIES AFTER ANOTHER THAT SOME OF THE ANTS WERE IN THE EARLY STAGES OF INVADING NEW GUINEA AND THE EASTERN MELANESIAN ARCHIPELAGOES.

OTHER SPECIES, APPARENTLY SURVIVORS FROM OLDER INVASIONS, WERE SPLITTING OFF AS FORMS LIMITED TO ONE ISLAND OR ANOTHER.

AND...AND...

AND STILL OTHER ENSEMBLES OF SPECIES WERE CLEARLY IN RETREAT, THEIR POPULATIONS SCATTERED HERE AND THERE IN POCKETS OF ISLAND TERRAIN. FINALLY, A SMALL PERCENTAGE HAD BEGUN TO EXPAND AGAIN, THIS TIME FROM NEW GUINEA.
IT CAME WITHIN A FEW MINUTES ONE JANUARY MORNING IN 1959.
THIS WHOLE CYCLE IS A MICROCOSM OF THE WORLDWIDE CYCLE ENVISIONED BY MATTHEW AND DARLINGTON.
THE PATTERN IS OBVIOUS, THE ONLY ONE POSSIBLE.
TO FIND THE SAME BIOGEOGRAPHIC PATTERN IN MINIATURE WAS A SURPRISE THEN, ALTHOUGH IN RETROSPECT, IT SEEMS ALMOST SELF-EVIDENT.
THE EXPANDING, DOMINANT SPECIES ARE ADAPTED FOR ECOLOGICALLY MARGINAL HABITATS—SAVANNAS, THE MONSOON FORESTS, THE SALT-LASHED BEACHES—IN WHICH RELATIVELY SMALL NUMBERS OF ANT SPECIES OCCUR.

I WAS TOO PLEASED WITH MYSELF TO WORRY.

IN TWO ARTICLES, I REFINED MY ANALYSIS. THE EXPANDING SPECIES, I REPORTED, HAVE CERTAIN CHARACTERISTICS ASSOCIATED WITH LIFE IN THE MARGINAL HABITATS.

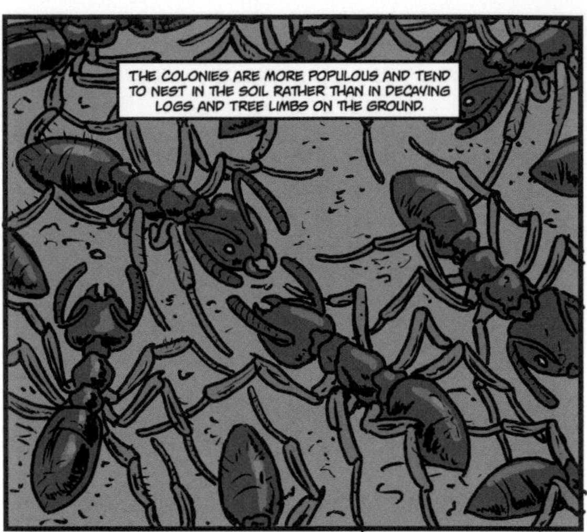
THE COLONIES ARE MORE POPULOUS AND TEND TO NEST IN THE SOIL RATHER THAN IN DECAYING LOGS AND TREE LIMBS ON THE GROUND.

THE WORKERS POSSESS MORE SPINES ON THE BODY, AN ARMAMENT USED AGAINST ENEMIES IN THE OPEN SPACES OF THE MARGINAL HABITATS.

THEY ORIENT MORE FREQUENTLY BY ODOR TRAILS LAID BY SCOUTS OVER THE GROUND.

THESE TRAITS ARE NOT THE SOURCE OF THE DOMINANCE, HOWEVER THEY ARE ONLY ADAPTATIONS TO LIFE IN THE MARGINAL HABITATS.

AND THAT'S THE SECOND ARTICLE—"THE NATURE OF THE TAXON CYCLE IN THE MELANESIAN ANT FAUNA."
"TAXON CYCLE" IS MY NAME FOR IT, YOU SEE.
WELL, AS TITLES GO IT'S NO "A MIDSUMMER NIGHT'S DREAM," BUT...

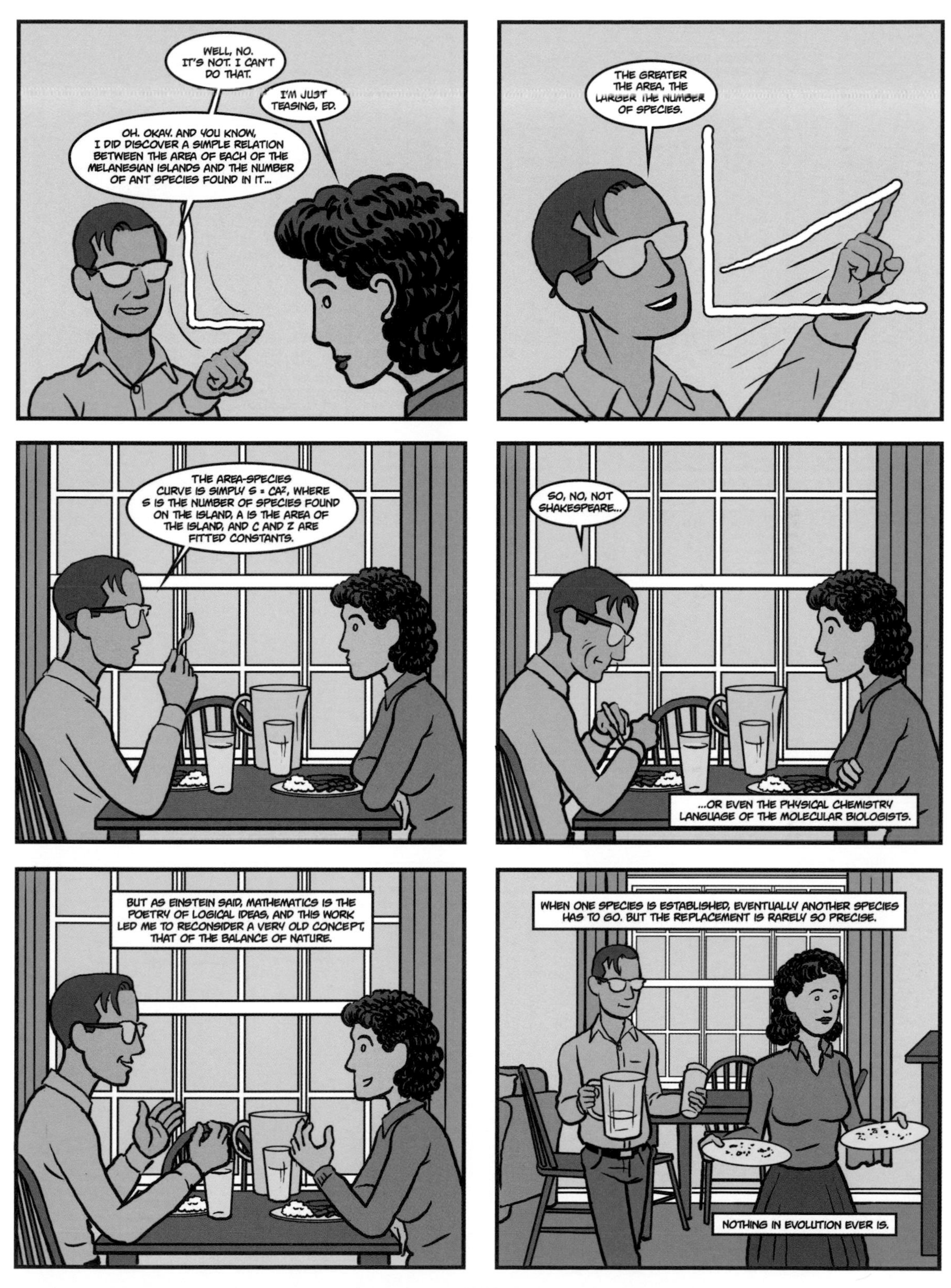
WELL, NO. IT'S NOT. I CAN'T DO THAT.
I'M JUST TEASING, ED.
OH. OKAY. AND YOU KNOW, I DID DISCOVER A SIMPLE RELATION BETWEEN THE AREA OF EACH OF THE MELANESIAN ISLANDS AND THE NUMBER OF ANT SPECIES FOUND IN IT...
THE GREATER THE AREA, THE LARGER THE NUMBER OF SPECIES.
THE AREA-SPECIES CURVE IS SIMPLY $S = CA^Z$, WHERE S IS THE NUMBER OF SPECIES FOUND ON THE ISLAND, A IS THE AREA OF THE ISLAND, AND C AND Z ARE FITTED CONSTANTS.
SO, NO, NOT SHAKESPEARE...
...OR EVEN THE PHYSICAL CHEMISTRY LANGUAGE OF THE MOLECULAR BIOLOGISTS.
BUT AS EINSTEIN SAID, MATHEMATICS IS THE POETRY OF LOGICAL IDEAS, AND THIS WORK LED ME TO RECONSIDER A VERY OLD CONCEPT, THAT OF THE BALANCE OF NATURE.
WHEN ONE SPECIES IS ESTABLISHED, EVENTUALLY ANOTHER SPECIES HAS TO GO. BUT THE REPLACEMENT IS RARELY SO PRECISE.
NOTHING IN EVOLUTION EVER IS.

HAPPENSTANCE DRIVES THE HISTORY OF FAUNA.
LIKE THE PEOPLE OF SOME ISLAND CIVILIZATIONS, SOME ANT SPECIES ACHIEVE DOMINANCE SIMPLY BY THEIR ABILITY TO CROSS THE SEA.
THERE WAS A BOOK'S WORTH OF MATERIAL ON POPULATION BIOLOGY, AND I WANTED TO WRITE IT. BUT BY 1961, THE TROPICS HAD REASSERTED THEIR PULL.
IN FEBRUARY, RENEE AND I TRAVELED TO TRINIDAD WHERE WE STAYED AS GUESTS OF ASA WRIGHT AT HER PROPERTY, SPRING HILL ESTATE, PERCHED NEAR THE HEAD OF ARIMA VALLEY, IN THE NORTH RANGE.
LOVELY PLACE.
WE SPENT SOME TIME IN SURINAME, TO ADD FIELD WORK IN THE FORESTS OF ZANDERIJ, THEN RETURNED TO SPRING HILL FOR A WHILE BEFORE PROCEEDING TO TOBAGO FOR THE FINAL THREE MONTHS OF OUR TOUR.

I FELT COMPLETELY AT HOME AGAIN IN THE HEAT AND SMELL OF ROTTING VEGETATION. AS FOR RENEE...
...SO MUCH INTERESTING FAUNA HERE. AMAZING DIVERSITY. FOR INSTANCE...

ED, WHAT'S THIS?
AH. GOOD EYE. THAT'S FROM THE VAMPIRE BATS.

WHAT KIND OF BATS?
OOPS

WELL...IT'S JUST A NAME. AND REGARDLESS, THEY ONLY FEED AFTER DARK, AND WE'LL BE BACK WELL BEFORE THE SUN GOES DOWN.
YOU'RE DARN RIGHT WE WILL, ED WILSON.
SHE DID NOT FEEL COMPLETELY AT HOME.

DACETON ARMIGERUM
DISCOVERIES CAME EASILY, AS ALWAYS FOR ME IN TROPICAL FORESTS. IN SURINAME, I ACQUIRED A COLONY OF THE GIANT, PRIMITIVE DACETINE ANT FROM A NEST HIGH IN A TREE AND MADE THE FIRST STUDY OF ITS SOCIAL ORGANIZATION.

I REDISCOVERED THE APPARENTLY "TRUE" CAVE ANT IN A CENTRAL TRINIDAD CAVERN.
SPELAEOMYRMEX URICHI!

I ALSO PROVED THAT THE SPECIES ALSO LIVES IN THE OPEN FOREST OF SURINAME—AND THUS IS NOT AN OBLIGATORY CAVE ANT.
LOOK! IT'S SPELAEOMYRMEX URICHI AGAIN

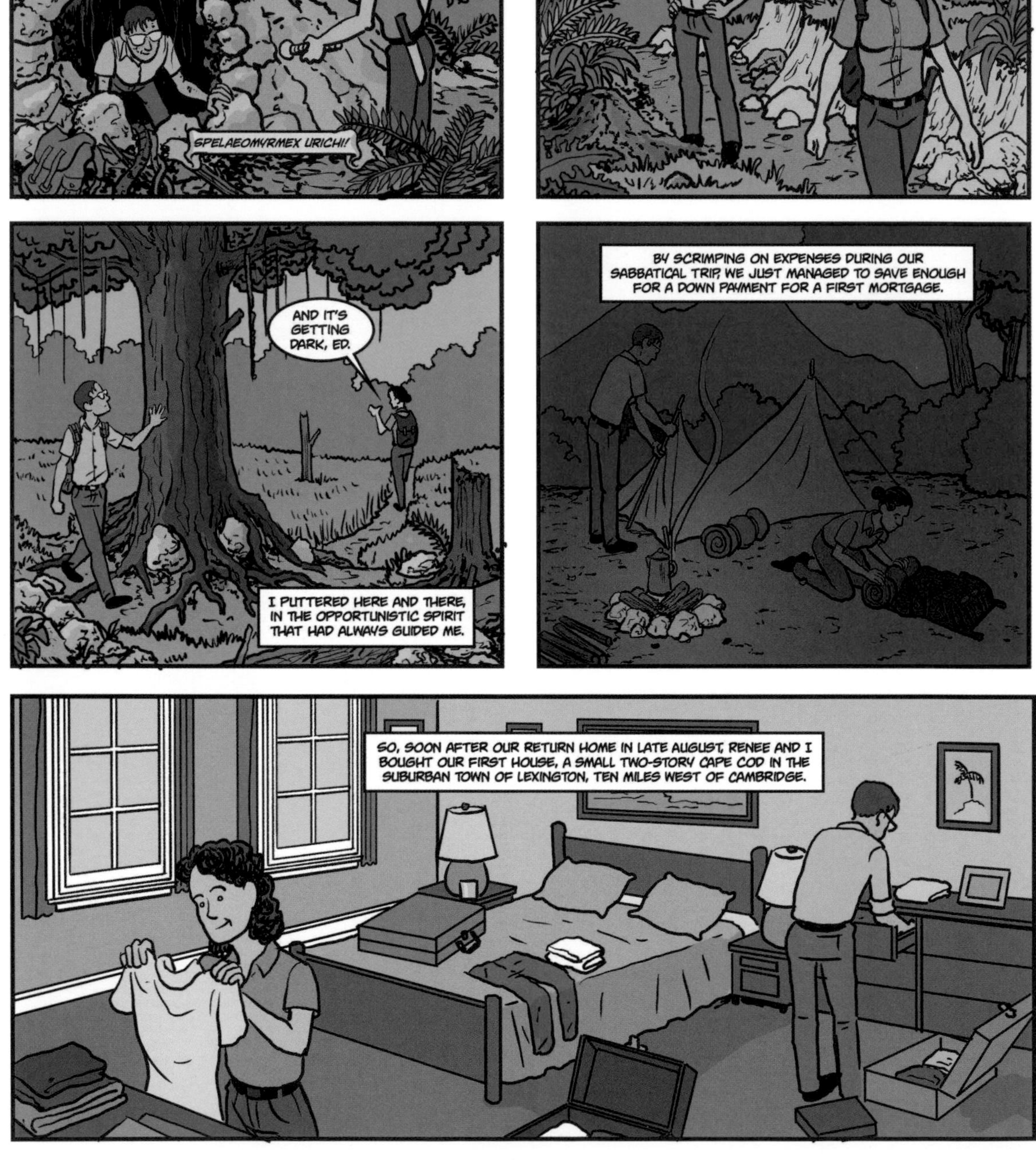
AND IT'S GETTING DARK, ED.
I PUTTERED HERE AND THERE, IN THE OPPORTUNISTIC SPIRIT THAT HAD ALWAYS GUIDED ME.
BY SCRIMPING ON EXPENSES DURING OUR SABBATICAL TRIP, WE JUST MANAGED TO SAVE ENOUGH FOR A DOWN PAYMENT FOR A FIRST MORTGAGE.
SO, SOON AFTER OUR RETURN HOME IN LATE AUGUST, RENEE AND I BOUGHT OUR FIRST HOUSE, A SMALL TWO-STORY CAPE COD IN THE SUBURBAN TOWN OF LEXINGTON, TEN MILES WEST OF CAMBRIDGE.

NOW FIVE YEARS INTO OUR MARRIAGE, WE AT LAST FELT ROOTED AND SECURE.

I FELT MORE CONFIDENT IN MY WORK AND IN THE KNOWLEDGE THAT I WOULD PROBABLY STAY AT HARVARD FOR THE REMAINDER OF MY CAREER.

DO YOU NEED TO GET BACK TO THE LAB?
WELL...

I WAS READY TO WRITE THAT BOOK ON POPULATION BIOLOGY.

FOR THIS PROJECT, I WAS JOINED BY ROBERT MACARTHUR.
ROBERT'S A REAL THEORETICIAN.
WELL...
I THINK WE NEED SOMEONE ELSE CLOSER TO PURE THEORY, WITH A BETTER MATHEMATICAL BACKGROUND.

THAT'S HOW WHAT I CALL THE TAXON CYCLE RELATES TO THE BALANCE OF SPECIES.

BIOGEOGRAPHY IS STILL LARGELY DESCRIPTIVE. THE MOST INTERESTING THEORY IS THE MATTHEW-DARLINGTON CYCLE OF DOMINANCE AND REPLACEMENT.
TIME FOR SOME NEW THINKING, THEN.
THE BEST SCIENCE COMES FROM INVENTING NEW CLASSIFICATIONS OF NATURAL PHENOMENA, ONES THAT SUGGEST HYPOTHESES AND NEW ROUNDS OF DATA GATHERING.
"ART IS THE LIE THAT HELPS US SEE THE TRUTH."
THAT'S PICASSO.
SURE, AND IT'S A NICE QUOTE. BUT LYING?
?
WELL, IT'S A METAPHOR. JUST LIKE WHEN WE CALL NATURE A TAPESTRY.
OKAY, IF YOU LOOK AT IT THAT WAY.
ROBERT LOOKED AT NATURE INDEPENDENT FROM WHAT OTHERS THOUGHT OF IT, OR OF.
FOLLOWING HIS LEAD, I EXPRESSED THREE CONVICTIONS OF MY OWN.
ISLANDS
FIRST, ISLANDS ARE THE KEY TO RAPID PROGRESS IN BIOGEOGRAPHY. THE COMMUNITIES THEY CONTAIN ARE DISCRETE UNITS THAT ARE ISOLATED BY THE SEA AND CAN BE STUDIED IN MULTIPLES.
ISLANDS
POPULATION BIOLOGY
SECOND, ALL BIOGEOGRAPHY, INCLUDING THE HISTORIES OF FAUNAS AND FLORAS, CAN BE MADE A BRANCH OF POPULATION BIOLOGY.

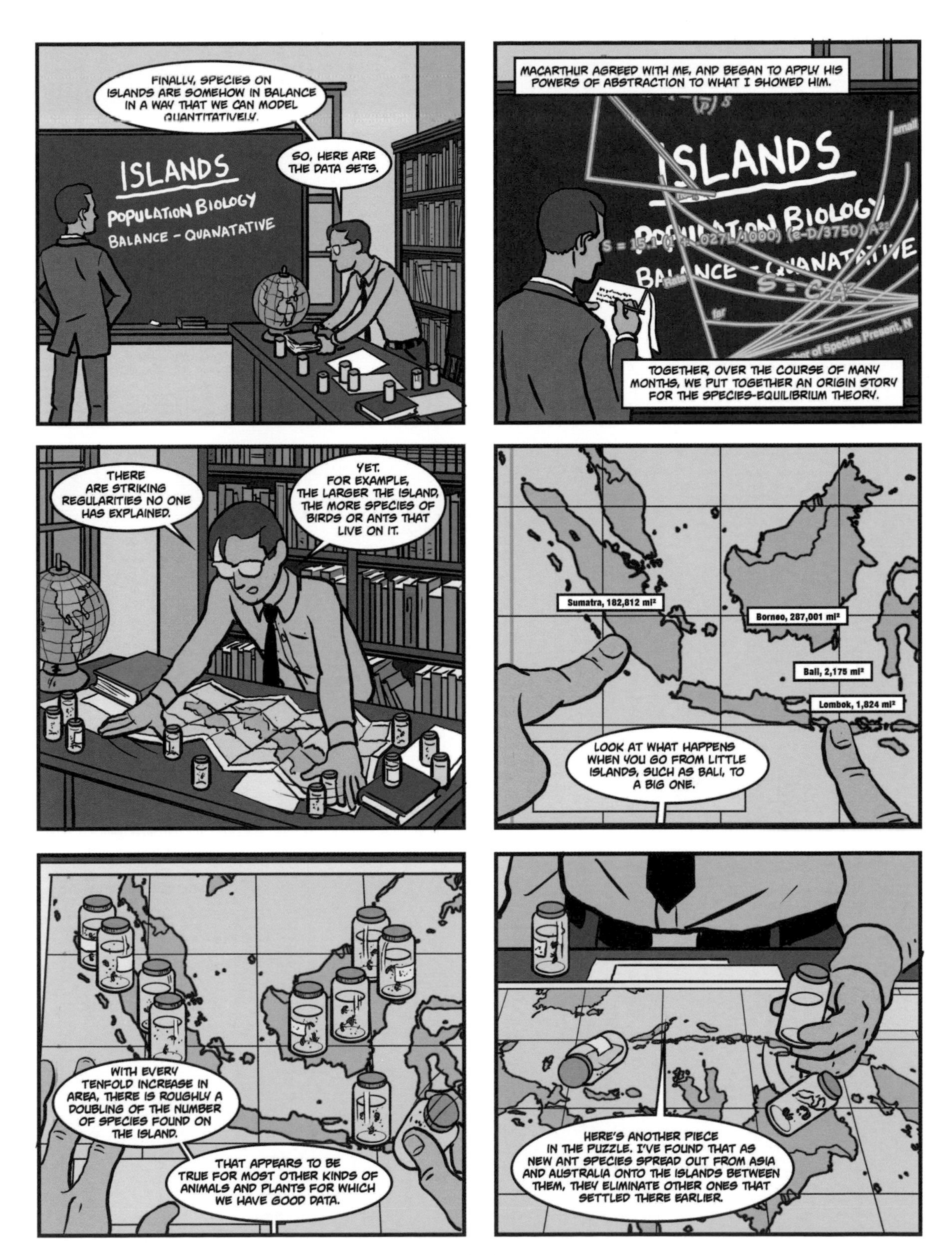

FINALLY, SPECIES ON ISLANDS ARE SOMEHOW IN BALANCE IN A WAY THAT WE CAN MODEL QUANTITATIVELY.
SO, HERE ARE THE DATA SETS.
ISLANDS
POPULATION BIOLOGY
BALANCE - QUANATATIVE
MACARTHUR AGREED WITH ME, AND BEGAN TO APPLY HIS POWERS OF ABSTRACTION TO WHAT I SHOWED HIM.
TOGETHER, OVER THE COURSE OF MANY MONTHS, WE PUT TOGETHER AN ORIGIN STORY FOR THE SPECIES-EQUILIBRIUM THEORY.
THERE ARE STRIKING REGULARITIES NO ONE HAS EXPLAINED.
YET. FOR EXAMPLE, THE LARGER THE ISLAND, THE MORE SPECIES OF BIRDS OR ANTS THAT LIVE ON IT.
Sumatra, 182,812 mi^2
Borneo, 287,001 mi^2
Bali, 2,175 mi^2
Lombok, 1,824 mi^2
LOOK AT WHAT HAPPENS WHEN YOU GO FROM LITTLE ISLANDS, SUCH AS BALI, TO A BIG ONE.
WITH EVERY TENFOLD INCREASE IN AREA, THERE IS ROUGHLY A DOUBLING OF THE NUMBER OF SPECIES FOUND ON THE ISLAND.
THAT APPEARS TO BE TRUE FOR MOST OTHER KINDS OF ANIMALS AND PLANTS FOR WHICH WE HAVE GOOD DATA.
HERE'S ANOTHER PIECE IN THE PUZZLE. I'VE FOUND THAT AS NEW ANT SPECIES SPREAD OUT FROM ASIA AND AUSTRALIA ONTO THE ISLANDS BETWEEN THEM, THEY ELIMINATE OTHER ONES THAT SETTLED THERE EARLIER.

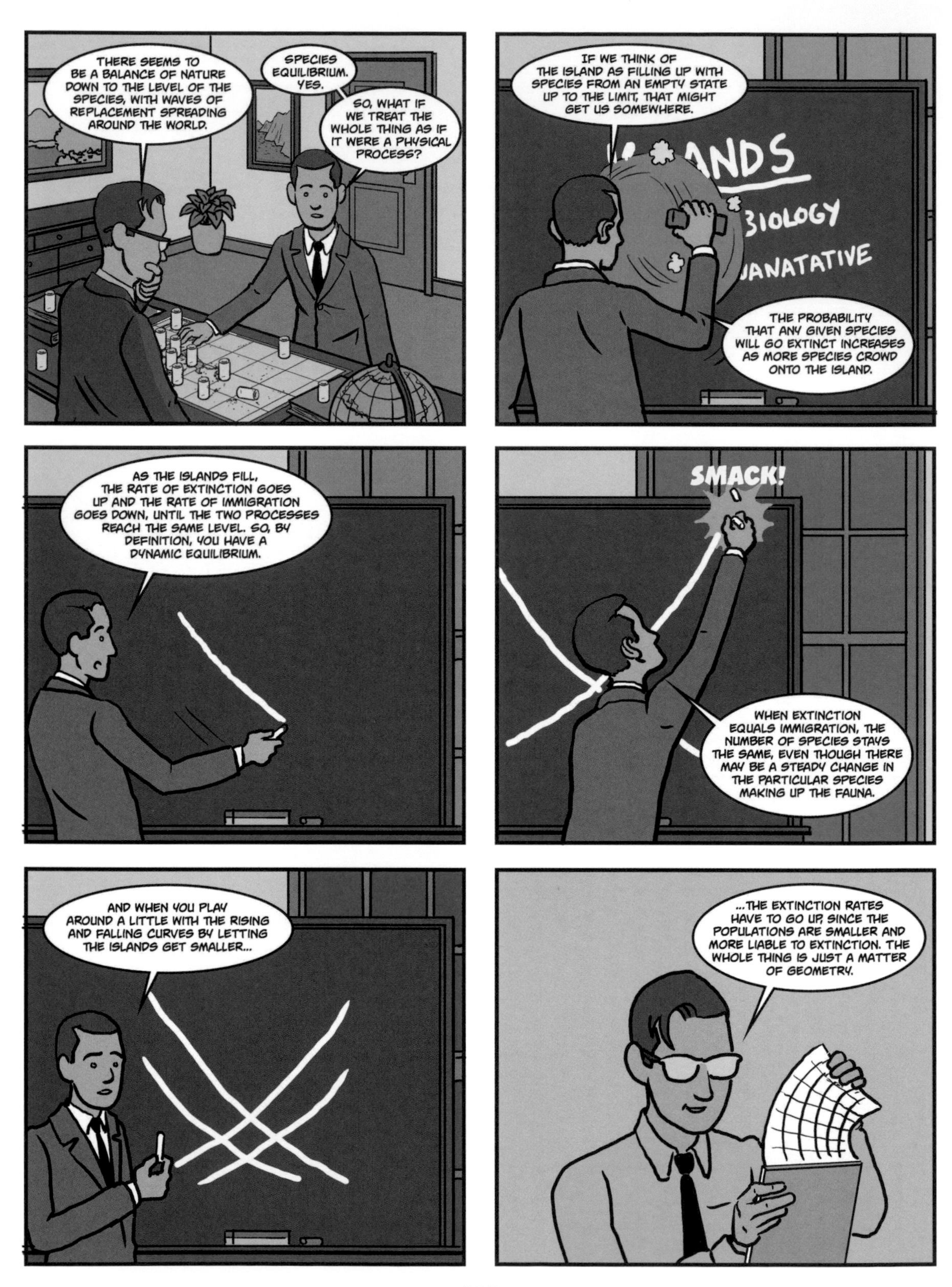
THERE SEEMS TO BE A BALANCE OF NATURE DOWN TO THE LEVEL OF THE SPECIES, WITH WAVES OF REPLACEMENT SPREADING AROUND THE WORLD.
SPECIES EQUILIBRIUM. YES.
SO, WHAT IF WE TREAT THE WHOLE THING AS IF IT WERE A PHYSICAL PROCESS?
IF WE THINK OF THE ISLAND AS FILLING UP WITH SPECIES FROM AN EMPTY STATE UP TO THE LIMIT, THAT MIGHT GET US SOMEWHERE.
THE PROBABILITY THAT ANY GIVEN SPECIES WILL GO EXTINCT INCREASES AS MORE SPECIES CROWD ONTO THE ISLAND.
AS THE ISLANDS FILL, THE RATE OF EXTINCTION GOES UP AND THE RATE OF IMMIGRATION GOES DOWN, UNTIL THE TWO PROCESSES REACH THE SAME LEVEL. SO, BY DEFINITION, YOU HAVE A DYNAMIC EQUILIBRIUM.
SMACK!
WHEN EXTINCTION EQUALS IMMIGRATION, THE NUMBER OF SPECIES STAYS THE SAME, EVEN THOUGH THERE MAY BE A STEADY CHANGE IN THE PARTICULAR SPECIES MAKING UP THE FAUNA.
AND WHEN YOU PLAY AROUND A LITTLE WITH THE RISING AND FALLING CURVES BY LETTING THE ISLANDS GET SMALLER...
...THE EXTINCTION RATES HAVE TO GO UP, SINCE THE POPULATIONS ARE SMALLER AND MORE LIABLE TO EXTINCTION. THE WHOLE THING IS JUST A MATTER OF GEOMETRY.

SO FAR SO GOOD. THE NUMBERS OF BIRD AND ANT SPECIES DO GO DOWN AS ISLANDS GET SMALLER AND FARTHER FROM THE MAINLAND. WE'LL LABEL THE TWO TRENDS *THE AREA EFFECT* AND *THE DISTANCE EFFECT*.

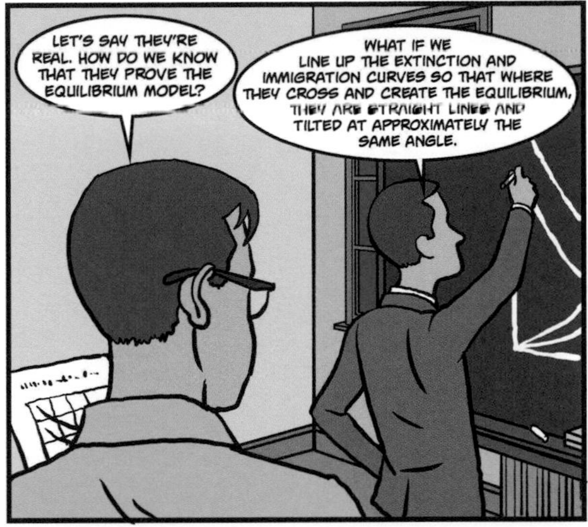
LET'S SAY THEY'RE REAL. HOW DO WE KNOW THAT THEY PROVE THE EQUILIBRIUM MODEL?
WHAT IF WE LINE UP THE EXTINCTION AND IMMIGRATION CURVES SO THAT WHERE THEY CROSS AND CREATE THE EQUILIBRIUM, THEY ARE STRAIGHT LINES AND TILTED AT APPROXIMATELY THE SAME ANGLE.

SOME ELEMENTARY DIFFERENTIAL CALCULUS...
UH-OH.
...AND WE CAN SHOW THAT THE NUMBER OF YEARS AN ISLAND TAKES TO FILL UP TO 90% OF ITS POTENTIAL SHOULD JUST ABOUT EQUAL THE NUMBER OF SPECIES AT EQUILIBRIUM DIVIDED BY THE NUMBER GOING EXTINCT EVERY YEAR.

"SHOULD..."
$-\frac{\ln 2}{20}$

WE CAN TEST THAT. THINK BACK ALMOST 100 YEARS.

KRAKATOA.

KRAKATOA
AUGUST 26, 1883
KRAKATOA
AUGUST 27, 1883
KRAKATOA
AUGUST 28, 1883
KRAKATOA (OR KRAKATAU), A SMALL ISLAND BETWEEN SUMATRA AND JAVA, GOT WIPED CLEAN OF ALL LIFE IN A SINGLE DAY.
ECOLOGISTS MANAGED TO KEEP A SPOTTY BUT SERVICEABLE RECORD OF THE RETURN OF BIRDS, PLANTS, AND A FEW OTHER ORGANISMS TO THE BARE VOLCANIC SLOPES.

OUR BASIC EQUILIBRIUM MODEL PREDICTED THAT THE BIRDS FOR WHICH THE BEST DATA WERE AVAILABLE, SHOULD REACH EQUILIBRIUM AT ABOUT THIRTY SPECIES.
UPON APPROACHING THAT LEVEL, THE FAUNA SHOULD BE LOSING ONE ESTABLISHED SPECIES BY LOCAL EXTINCTION EACH YEAR WHILE ACQUIRING ONE NEW SPECIES BY IMMIGRATION.
UNFORTUNATELY, THE DATA INDICATED THAT WHILE BIRDS INDEED APPEAR TO BE LEVELING OFF AT APPROXIMATELY THIRTY SPECIES, THE TURNOVER WAS ONE SPECIES EVERY FIVE YEARS, NOT ONE EACH YEAR.
IS OUR MODEL REALLY OFF FIVEFOLD, OR IS THE DISCREPANCY DUE TO SAMPLING ERRORS WAY BACK WHEN?
NO WAY TO TELL.
WE NEEDED MORE RELIABLE DATA SETS TO ADVANCE EQUILIBRIUM THEORY IN A SERIOUS WAY.

UNFORTUNATELY, WE CAN'T MAKE A VOLCANO ERUPT AND WIPE OUT AN ISLAND ON COMMAND, SO WE'RE GOING TO HAVE TO...
EXCUSE ME?

I SAID "UNFORTUNATELY"!
BUT I HAVE AN IDEA—AN INTERESTING TWIST ON FIELD WORK.

ED...
NOT FAR THIS TIME! I'VE GOT THE EXPERIMENTS TO DO HERE ON THE SOCIAL BEHAVIOR OF ANTS, AND THERE'S US.

AND YOU YOU YOU! WHO'S MY BEST GIRL?
PUELLA OPTIMUS (CATHERINE)

SO, I'M THINKING CLOSER TO HOME.

MY UNUSUAL, RATHER BIZARRE IDEA IS TO SET UP AN EXPERIMENTAL SYSTEM IN AN ARCHIPELAGO AND CREATE KRAKATOAS AT WILL.
AND THEN, WATCH THEIR RECOLONIZATION AT LEISURE.

I'D NEED A LOT OF KRAKATOAS, SO THERE'S ONLY ONE WAY TO SOLVE THE PROBLEM—MINIATURIZE THE SYSTEM!
SO, INSTEAD OF RELYING ON CONVENTIONAL ISLANDS, WHY NOT USE TINY ONES, AT MOST A FEW HUNDRED SQUARE METERS?
NO MAMMALS, BIRDS, OR ANY OTHER LAND VERTEBRATES ABOVE THE SIZE OF SMALL LIZARDS, SO, BIOLOGISTS WOULD NOT CALL THEM ISLANDS AT ALL.
BUT BUGS? TO AN ANT OR SPIDER, A SINGLE TREE IS LIKE A WHOLE FOREST. THE LIFETIME OF SUCH A CREATURE CAN BE SPENT IN A MICROTERRITORY THE SIZE OF A DINNER PLATE.
MAINE
FLORIDA
THERE ARE ISLANDS OF ALL SIZES AND TYPES RIGHT HERE IN THE U.S.
I PREFER MARINE WATERS OVER LAKES AND RIVERS.
WHY?
UMM. I...YOU KNOW, I GUESS IT'S AN AESTHETIC CHOICE.

A DECISIVE WINNER QUICKLY EMERGED: THE FLORIDA KEYS SEEMED IDEAL. THE ISLANDS CAME IN ALL SIZES, FROM SINGLE TREES TO SIZABLE EXPANSES UP TO A SQUARE KILOMETER OR MORE.
Big Pine Key
Marathon
Key West
83°W
82°W
Curacao
THEY VARIED IN DEGREES OF ISOLATION FROM A FEW METERS TO HUNDREDS OF METERS FROM THE NEAREST NEIGHBOR.
Big Pine Key
Marathon
81° W
THE FORESTS ON THEM WERE SIMPLE, CONSISTING IN MOST CASES ENTIRELY OF RED MANGROVE TREES. AND THEY WERE AVAILABLE IN VAST NUMBERS.
RHIZOPHORA MANGLE
AND I COULD BE AT MY ISLANDS IN A SINGLE DAY...
...IF I STARTED WITH AN EARLY FOUR-HOUR FLIGHT FROM BOSTON TO MIAMI, DROVE A RENTAL CAR DOWN U.S. 1...

...AND FINALLY TOOK A SHORT BOAT TRIP OUT TO THE ISLANDS I HAD IN MIND.
IN JUNE 1965, RENEE AND CATHY JOINED ME ON A TRIP TO MIAMI AND MY NEW ISLAND WORLD.
MOTEL
AHH. BACK WHERE I WAS MEANT TO BE!
HERE?
WELL, NOT PRECISELY HERE...
FEW PEOPLE VENTURED INTO THE SWAMPY ARCHIPELAGOES OF MY CHOICE, LESS THAN A MILE AWAY FROM U.S. 1.
ONCE OR TWICE A DAY, I SAW A DISTANT FISHERMAN OR A POWERBOAT OUT IN DEEPER WATER, BUT BEYOND HEARING RANGE OF THE RUMBLE AND WHINE OF TRAFFIC, THE SWAMPS AND ISLETS WERE PRISTINE, A VIRGIN WILDERNESS.
PASS THE TANNING OIL, BECKY.
RUMBLE!
NO ONE BUT A NATURALIST—OR ESCAPED CONVICT—WOULD CHOOSE TO TRAVERSE THE GLUE-LIKE MUD FLATS AND CLIMB THROUGH THE TANGLED PROP ROOTS AND TRUNKS OF THE MANGROVE TREES.
I HAD IT ALL TO MYSELF—ONE MORE TIME...
A WORLD MORE COMPLEX AND BEAUTIFUL THAN ANYTHING CONTRIVED BY HUMAN ENTERPRISE.

MANY SPECIES FLOURISHED IN BREEDING POPULATIONS, CRUCIAL FOR THE EXPERIMENTAL BIOGEOGRAPHY I HAD IN MIND. AND FROM ONE MANGROVE CLUMP TO THE NEXT, THE SPECIES CHANGED.

FOR ANTS, THE PATTERN WAS CONSISTENT WITH COMPETITIVE EXCLUSION. BELOW A CERTAIN ISLAND SIZE, THE COLONIZATION OF SOME SPECIES APPEARED TO PRECLUDE THE ESTABLISHMENT OF OTHERS.

I CAN ANALYZE THE DISTRIBUTION OF ARTHROPODS IN JUST DAYS OR WEEKS. NOT YEARS, LIKE OUT IN THE PACIFIC, LOOKING AT BIRDS.
SOUNDS GOOD TO ME.
IN ADDITION, I CAN SELECT OTHER ISLANDS LACKING TREES IN ORDER TO MAKE STERILIZATION EASIER.

IF I CAN MONITOR THEM BEFORE AND AFTER A BIG STORM, I MIGHT OBSERVE THE RECOLONIZATION PROCESS AND ESTABLISH WHETHER IT CREATED AN EQUILIBRIUM.
LET THE CARIBBEAN'S STORMY WEATHER BE THE VOLCANO.

AND LUCKILY...
ED!

PROVIDENTIALLY—FROM A BIOLOGIST'S *POSSIBLY* PERVERSE POINT OF VIEW—TWO HURRICANES STRUCK BETWEEN SEPTEMBER 1965 AND JUNE 1966.
BETSY
125 MPH WINDS

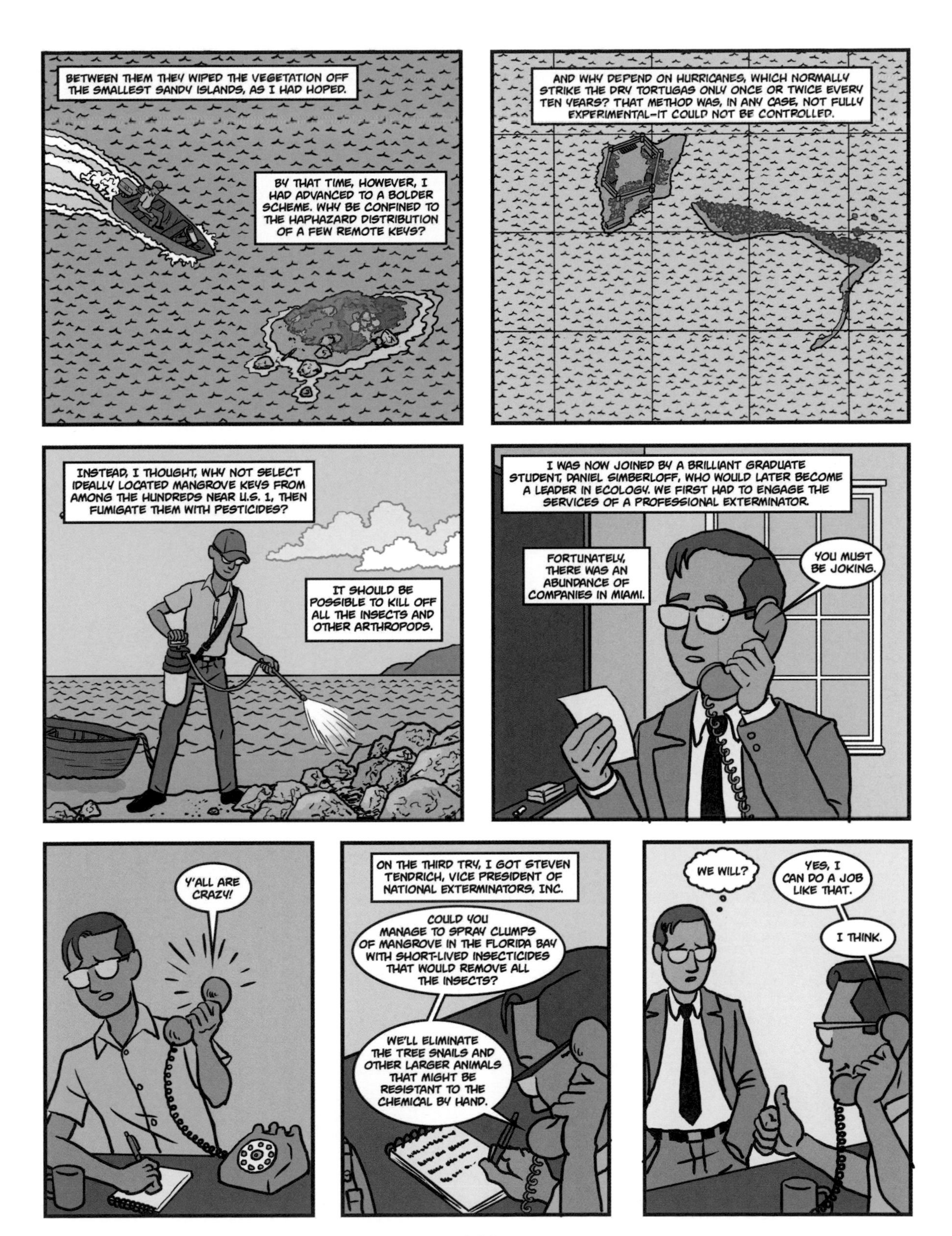
BETWEEN THEM THEY WIPED THE VEGETATION OFF THE SMALLEST SANDY ISLANDS, AS I HAD HOPED.
BY THAT TIME, HOWEVER, I HAD ADVANCED TO A BOLDER SCHEME. WHY BE CONFINED TO THE HAPHAZARD DISTRIBUTION OF A FEW REMOTE KEYS?
AND WHY DEPEND ON HURRICANES, WHICH NORMALLY STRIKE THE DRY TORTUGAS ONLY ONCE OR TWICE EVERY TEN YEARS? THAT METHOD WAS, IN ANY CASE, NOT FULLY EXPERIMENTAL—IT COULD NOT BE CONTROLLED.
INSTEAD, I THOUGHT, WHY NOT SELECT IDEALLY LOCATED MANGROVE KEYS FROM AMONG THE HUNDREDS NEAR U.S. 1, THEN FUMIGATE THEM WITH PESTICIDES?
IT SHOULD BE POSSIBLE TO KILL OFF ALL THE INSECTS AND OTHER ARTHROPODS.
I WAS NOW JOINED BY A BRILLIANT GRADUATE STUDENT, DANIEL SIMBERLOFF, WHO WOULD LATER BECOME A LEADER IN ECOLOGY. WE FIRST HAD TO ENGAGE THE SERVICES OF A PROFESSIONAL EXTERMINATOR.
FORTUNATELY, THERE WAS AN ABUNDANCE OF COMPANIES IN MIAMI.
YOU MUST BE JOKING.
Y'ALL ARE CRAZY!
ON THE THIRD TRY, I GOT STEVEN TENDRICH, VICE PRESIDENT OF NATIONAL EXTERMINATORS, INC.
COULD YOU MANAGE TO SPRAY CLUMPS OF MANGROVE IN THE FLORIDA BAY WITH SHORT-LIVED INSECTICIDES THAT WOULD REMOVE ALL THE INSECTS?
WE'LL ELIMINATE THE TREE SNAILS AND OTHER LARGER ANIMALS THAT MIGHT BE RESISTANT TO THE CHEMICAL BY HAND.
WE WILL?
YES, I CAN DO A JOB LIKE THAT.
I THINK.

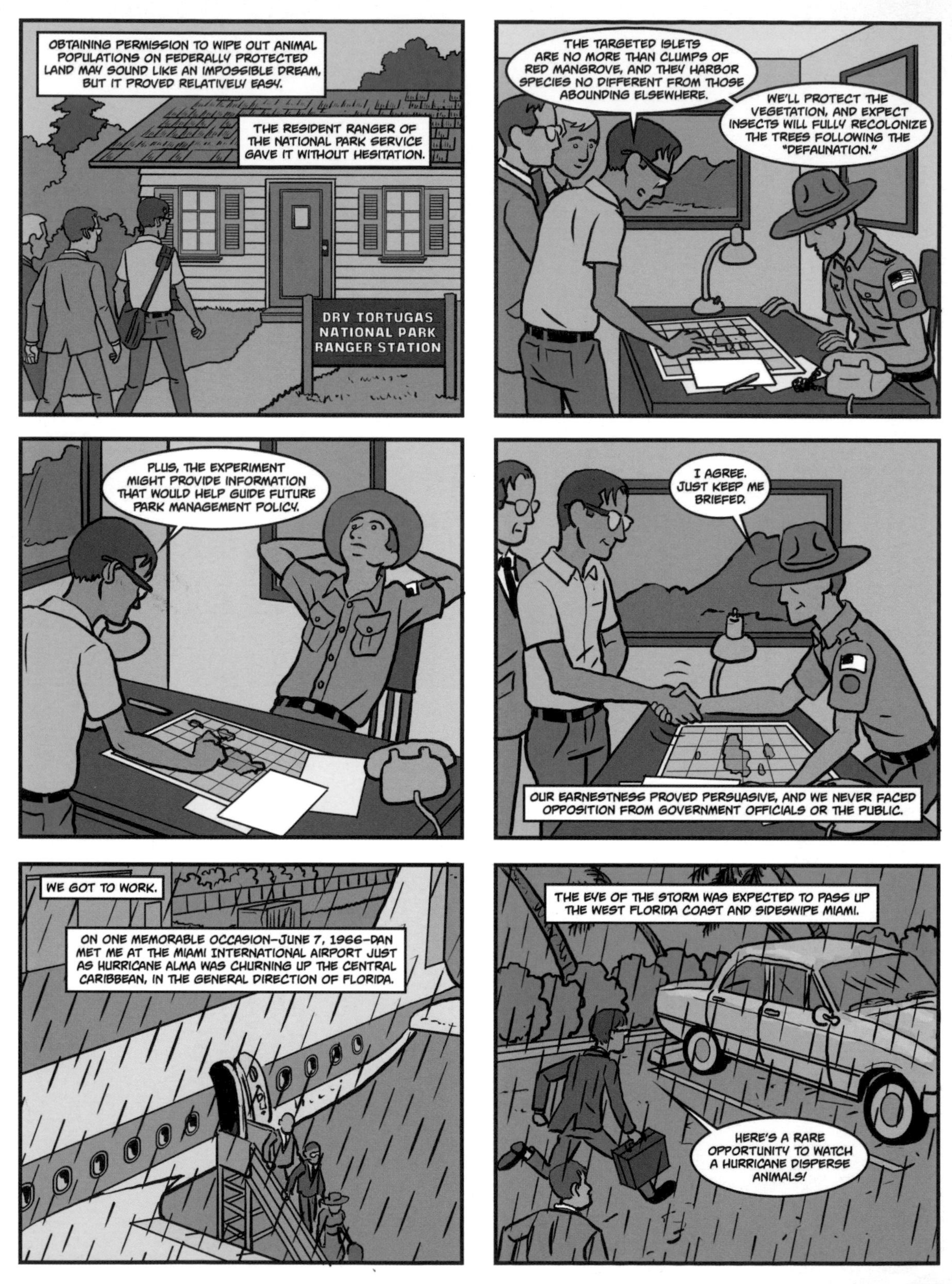
OBTAINING PERMISSION TO WIPE OUT ANIMAL POPULATIONS ON FEDERALLY PROTECTED LAND MAY SOUND LIKE AN IMPOSSIBLE DREAM, BUT IT PROVED RELATIVELY EASY.
THE RESIDENT RANGER OF THE NATIONAL PARK SERVICE GAVE IT WITHOUT HESITATION.
DRY TORTUGAS NATIONAL PARK RANGER STATION
THE TARGETED ISLETS ARE NO MORE THAN CLUMPS OF RED MANGROVE, AND THEY HARBOR SPECIES NO DIFFERENT FROM THOSE ABOUNDING ELSEWHERE.
WE'LL PROTECT THE VEGETATION, AND EXPECT INSECTS WILL FULLY RECOLONIZE THE TREES FOLLOWING THE "DEFAUNATION."
PLUS, THE EXPERIMENT MIGHT PROVIDE INFORMATION THAT WOULD HELP GUIDE FUTURE PARK MANAGEMENT POLICY.
I AGREE. JUST KEEP ME BRIEFED.
OUR EARNESTNESS PROVED PERSUASIVE, AND WE NEVER FACED OPPOSITION FROM GOVERNMENT OFFICIALS OR THE PUBLIC.
WE GOT TO WORK.
ON ONE MEMORABLE OCCASION—JUNE 7, 1966—DAN MET ME AT THE MIAMI INTERNATIONAL AIRPORT JUST AS HURRICANE ALMA WAS CHURNING UP THE CENTRAL CARIBBEAN, IN THE GENERAL DIRECTION OF FLORIDA.
THE EYE OF THE STORM WAS EXPECTED TO PASS UP THE WEST FLORIDA COAST AND SIDESWIPE MIAMI.
HERE'S A RARE OPPORTUNITY TO WATCH A HURRICANE DISPERSE ANIMALS!

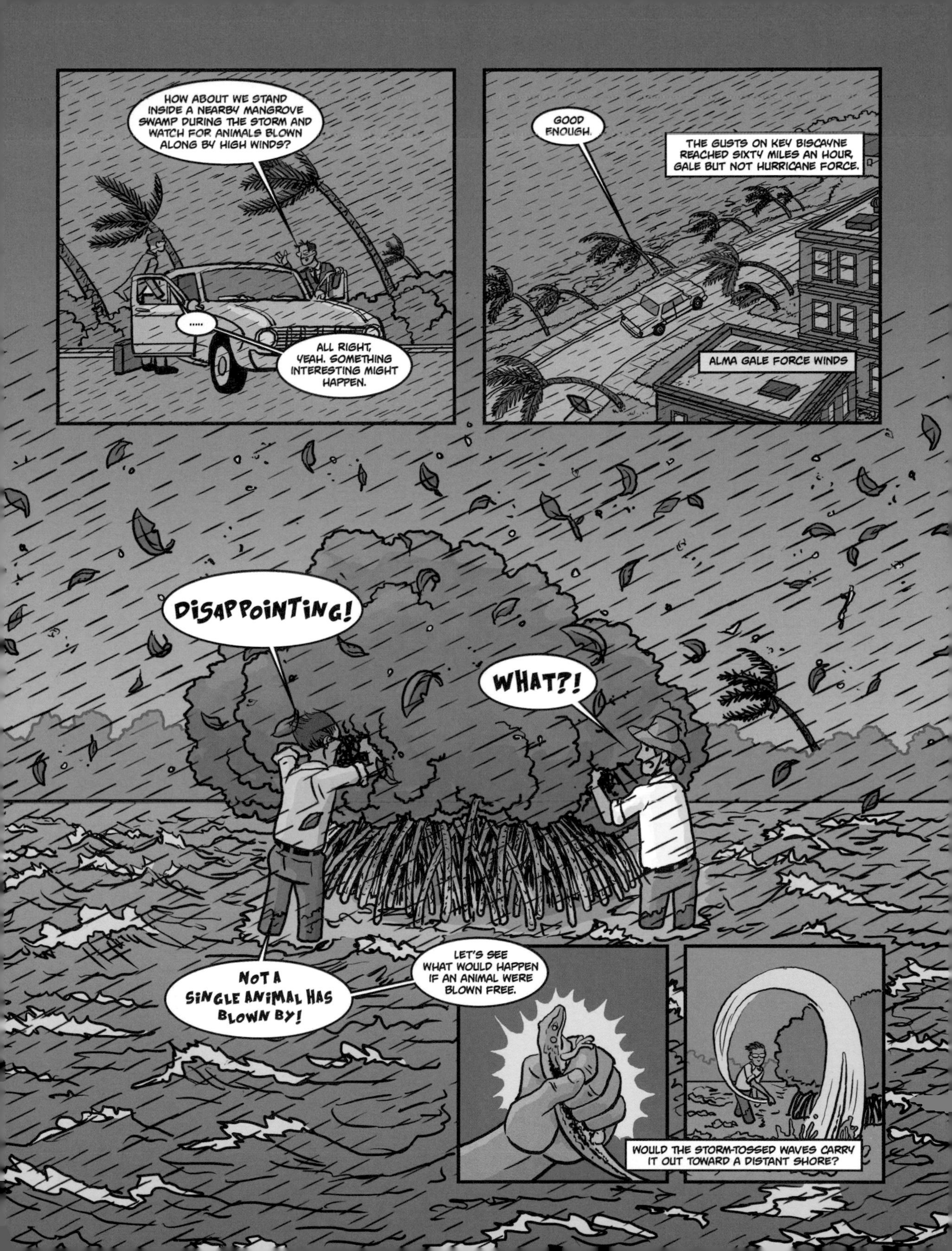
HOW ABOUT WE STAND INSIDE A NEARBY MANGROVE SWAMP DURING THE STORM AND WATCH FOR ANIMALS BLOWN ALONG BY HIGH WINDS?
.....
ALL RIGHT, YEAH. SOMETHING INTERESTING MIGHT HAPPEN.
GOOD ENOUGH.
THE GUSTS ON KEY BISCAYNE REACHED SIXTY MILES AN HOUR, GALE BUT NOT HURRICANE FORCE.
ALMA GALE FORCE WINDS
DISAPPOINTING!
WHAT?!
NOT A SINGLE ANIMAL HAS BLOWN BY!
LET'S SEE WHAT WOULD HAPPEN IF AN ANIMAL WERE BLOWN FREE.
WOULD THE STORM-TOSSED WAVES CARRY IT OUT TOWARD A DISTANT SHORE?

WE WERE BOTH A LITTLE CRAZY IN THOSE DAYS—IN LATER YEARS WE AGREED WE WERE LUCKY WE OURSELVES WEREN'T WASHED TO A DISTANT SHORE, PROVING OUR OWN HYPOTHESIS IN EXTREMIS.
A MONTH LATER, WE REACHED OUR FIRST LITTLE ISLAND AND SPRAYED IT WITH PARATHION.
ISLAND E1
THE NEXT MORNING, WE PROCEEDED TO E2.

UH-UH. NOPE.
THIS STUFF DON'T WORK ON SHARKS.
WE AIN'T GETTIN' PAID ENOUGH AS IT IS.
HUH? NURSE SHARKS NEVER ATTACK PEOPLE UNLESS HOOKED OR SEIZED BY THE TAIL. AND THEY ONLY EAT SMALL BOTTOM-DWELLING ANIMALS.

IT'S OKAY, FELLOWS. I'LL STAND GUARD.
IMPRESSED BY MY SPECIOUS BRAVERY AND WITH THEIR MALE PRIDE CHALLENGED, THE CREW GOT INTO THE WATER AND SPRAYED E2.

AFTER THE DEFAUNATION, DAN MONITORED THE MANGROVE FAUNA.
IT WAS HARD AND UNCOMFORTABLE WORK, DEMANDING THE COMBINED SKILLS OF INSECT SYSTEMATIST, ROOFER, AND RESTAURANT HEALTH INSPECTOR.

SIMBERLOFF, THE CITY-BRED MATHEMATICIAN, ENDURED THE INSECT BITES AND LONELY HOURS IN THE HOT SUN I HAD PROMISED HIM.

ONCE, AFTER HIS OUTBOARD MOTOR FAILED, HE SPENT THE NIGHT ON ONE OF THE ISLETS, MANAGING TO ESCAPE ONLY WHEN HE HAILED A PASSING FISHERMAN THE FOLLOWING MORNING.

EXASPERATED WITH THE GLUE-LIKE MUD WE HAD TO WADE THROUGH, HE BUILT A PAIR OF FOOTPADS AND DRILLED HOLES IN THEM.
BASIC PHYSICS, ED. THESE WILL REDUCE SUCTION WHEN LIFTED!

I'D HATE TO SEE FULL SUCTION! DON'T PATENT THOSE "SIMBERLOFFS" JUST YET, DAN.
VERY FUNNY.

UNFORTUNATELY, SOME BEETLE LARVAE LIVING DEEP IN THE WOOD OF DEAD BRANCHES SURVIVED. WE REALIZED WE HAD NO WAY OF KNOWING WHAT OTHER CREATURES MIGHT STILL LIVE IN THESE DEEPER SPACES.
TO RUN A PROPER EXPERIMENT, WE HAD TO START WITH ISLETS SCOURGED OF ALL ANIMAL LIFE. NO EXCEPTIONS.
WE HAVE TO FUMIGATE THE ISLETS WITH A POISONOUS GAS, ONE THAT PENETRATES EVERY CRACK AND CREVICE.
CAN YOU DO IT?
WHY NOT? IN MIAMI WE COVER ENTIRE HOUSES WITH A RUBBERIZED NYLON TENT AND FUMIGATE THE INTERIOR.
THAT KILLS ALL THE TERMITES AND OTHER PESTS, NO MATTER HOW DEEPLY HIDDEN IN THE WOODWORK.
DOING THIS TO A LARGE OBJECT SURROUNDED BY WATER WILL BE TRICKY...
WE CAN'T JUST LAY THE COVER ON TOP OF THE FRAGILE BRANCHES.
SNAP!
CRACK!
SNAP!
AND SOMETHING ELSE— THE DOSAGE OF GAS MUST BE HIGH ENOUGH TO KILL ALL THE ANIMALS, BUT LOW ENOUGH TO LEAVE THE MANGROVES UNDAMAGED.
STUDYING A GHOST ISLAND OF DEAD WOOD AND FALLEN LEAVES WOULD HAVE NO MEANING.
I AGREE. AND WE PROMISED THE NATIONAL PARK SERVICE WE WOULD PRESERVE THE LIVE VEGETATION.
THE CREW ERECTED A SCAFFOLDING AROUND THE ISLET AS A FRAME FOR THE TENT, AND TENDRICH FOUND THE NARROW WINDOW BETWEEN INSECT KILL AND TREE KILL.

THE NEXT DAY, WE SEARCHED THE ISLET THOROUGHLY AND FOUND NO TRACE OF ANIMAL LIFE. EVEN THE DEEP-BORING INSECTS HAD BEEN KILLED. OUR COLONIZATION EXPERIMENT WAS UNDERWAY AT LAST.

TENDRICH, HOWEVER, WAS NOT SATISFIED. HE ASKED RALPH NEVINS, A STEEPLEJACK, WHETHER IT WAS POSSIBLE TO ERECT A SMALL TOWER IN THE MIDDLE OF A MANGROVE SWAMP...
SURE, WHY NOT?

WILL IT BE DIFFICULT?
DON'T THINK SO.

THE REST OF OUR ISLETS WERE FUMIGATED BENEATH TENTS WRAPPED AROUND THE GUY WIRES OF A TOWER RAISED BY NEVINS.

BEMBIDION CONTRACTUM
VEIGAIAIDAE
LIGIA EXOTICA
THE RECOLONIZATION BY ARTHROPOD SPECIES WAS ALREADY WELL UNDERWAY WITHIN WEEKS.
TACHYS OCCULATOR
WINGED ANT QUEENS, NEWLY INSEMINATED DURING THEIR NUPTIAL FLIGHTS, LANDED, SHED THEIR WINGS, AND STARTED COLONIES.
PENTACORA SPHACELATA
ACTINOPTERYX FUCICOLA
CORYLOPHIDAE
PODURIDAE
LABIDURA RIPARIA
LYCOSIDAE
SPIDERS CAME EARLY IN ABUNDANCE; SOME WERE WOLF SPIDERS THE SIZE OF SILVER DOLLARS. SINCE THERE HAD BEEN NO MAJOR STORMS, WE GUESSED THAT THEY CROSSED THE WATER USING BALLOONING.

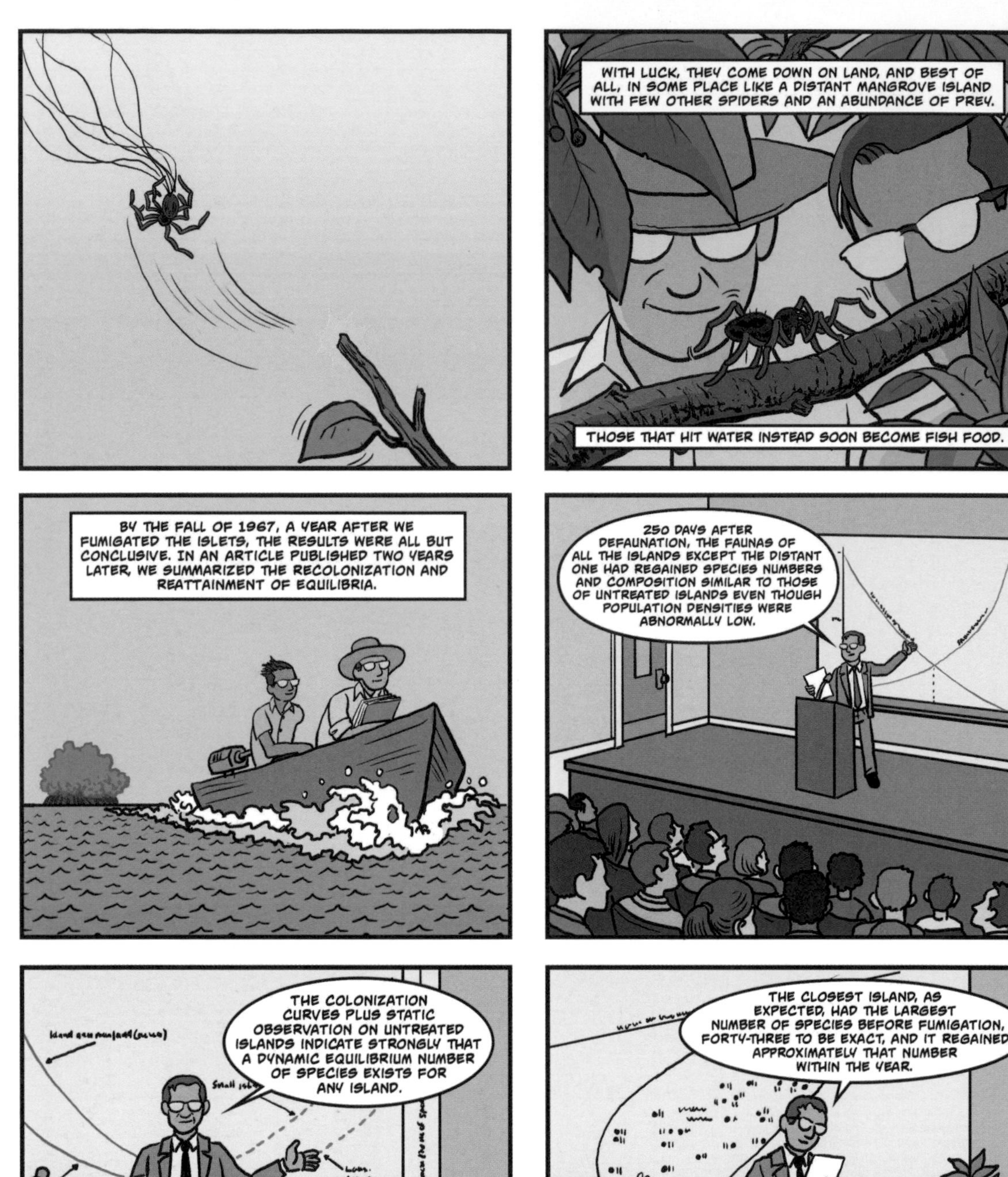
WITH LUCK, THEY COME DOWN ON LAND, AND BEST OF ALL, IN SOME PLACE LIKE A DISTANT MANGROVE ISLAND WITH FEW OTHER SPIDERS AND AN ABUNDANCE OF PREY.
THOSE THAT HIT WATER INSTEAD SOON BECOME FISH FOOD.
BY THE FALL OF 1967, A YEAR AFTER WE FUMIGATED THE ISLETS, THE RESULTS WERE ALL BUT CONCLUSIVE. IN AN ARTICLE PUBLISHED TWO YEARS LATER, WE SUMMARIZED THE RECOLONIZATION AND REATTAINMENT OF EQUILIBRIA.
250 DAYS AFTER DEFAUNATION, THE FAUNAS OF ALL THE ISLANDS EXCEPT THE DISTANT ONE HAD REGAINED SPECIES NUMBERS AND COMPOSITION SIMILAR TO THOSE OF UNTREATED ISLANDS EVEN THOUGH POPULATION DENSITIES WERE ABNORMALLY LOW.
THE COLONIZATION CURVES PLUS STATIC OBSERVATION ON UNTREATED ISLANDS INDICATE STRONGLY THAT A DYNAMIC EQUILIBRIUM NUMBER OF SPECIES EXISTS FOR ANY ISLAND.

THE CLOSEST ISLAND, AS EXPECTED, HAD THE LARGEST NUMBER OF SPECIES BEFORE FUMIGATION, FORTY-THREE TO BE EXACT, AND IT REGAINED APPROXIMATELY THAT NUMBER WITHIN THE YEAR.

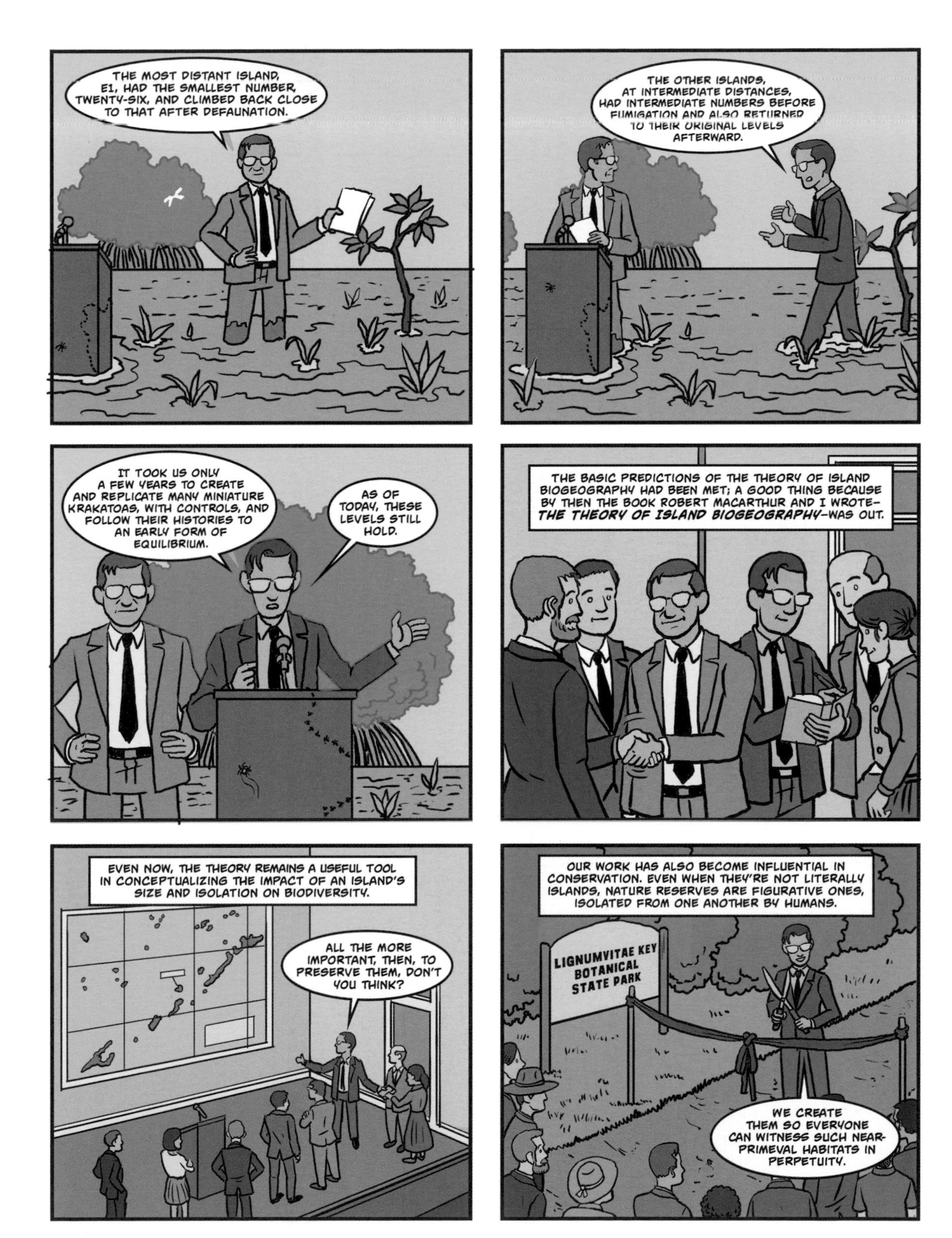

THE MOST DISTANT ISLAND, E1, HAD THE SMALLEST NUMBER, TWENTY-SIX, AND CLIMBED BACK CLOSE TO THAT AFTER DEFAUNATION.
THE OTHER ISLANDS, AT INTERMEDIATE DISTANCES, HAD INTERMEDIATE NUMBERS BEFORE FUMIGATION AND ALSO RETURNED TO THEIR ORIGINAL LEVELS AFTERWARD.
IT TOOK US ONLY A FEW YEARS TO CREATE AND REPLICATE MANY MINIATURE KRAKATOAS, WITH CONTROLS, AND FOLLOW THEIR HISTORIES TO AN EARLY FORM OF EQUILIBRIUM.
AS OF TODAY, THESE LEVELS STILL HOLD.
THE BASIC PREDICTIONS OF THE THEORY OF ISLAND BIOGEOGRAPHY HAD BEEN MET; A GOOD THING BECAUSE BY THEN THE BOOK ROBERT MACARTHUR AND I WROTE–*THE THEORY OF ISLAND BIOGEOGRAPHY*–WAS OUT.
EVEN NOW, THE THEORY REMAINS A USEFUL TOOL IN CONCEPTUALIZING THE IMPACT OF AN ISLAND'S SIZE AND ISOLATION ON BIODIVERSITY.
ALL THE MORE IMPORTANT, THEN, TO PRESERVE THEM, DON'T YOU THINK?
OUR WORK HAS ALSO BECOME INFLUENTIAL IN CONSERVATION. EVEN WHEN THEY'RE NOT LITERALLY ISLANDS, NATURE RESERVES ARE FIGURATIVE ONES, ISOLATED FROM ONE ANOTHER BY HUMANS.
LIGNUMVITAE KEY BOTANICAL STATE PARK
WE CREATE THEM SO EVERYONE CAN WITNESS SUCH NEAR-PRIMEVAL HABITATS IN PERPETUITY.

AND THEN I GOT FAMOUS. NOT FOR BIOGEOGRAPHY, BUT FOR...
WELL, YOU ALREADY GUESSED, I'M SURE.
THEY ARE EVERYWHERE, DARK AND RUDDY SPECKS THAT ZIGZAG ACROSS THE GROUND AND DOWN HOLES, MILLIGRAM-WEIGHT INHABITANTS OF AN ALIEN CIVILIZATION WHO HIDE THEIR DAILY ROUNDS FROM OUR EYES.
FOR OVER 50 MILLION YEARS, ANTS HAVE BEEN OVERWHELMINGLY DOMINANT INSECTS EVERYWHERE ON THE LAND OUTSIDE THE POLAR AND ALPINE ICE FIELDS.
BY MY ESTIMATE, BETWEEN 1 AND 10 MILLION BILLION INDIVIDUALS ARE ALIVE AT ANY MOMENT, ALL OF THEM TOGETHER WEIGHING, TO THE NEAREST ORDER OF MAGNITUDE, AS MUCH AS THE TOTALITY OF HUMAN BEINGS.
IF WE VANISHED, THE ENVIRONMENT WOULD RETURN TO THE FERTILE BALANCE THAT EXISTED BEFORE THE HUMAN POPULATION EXPLOSION.
ONLY A DOZEN OR SO SPECIES—THE CRAB LOUSE, FOR INSTANCE—DEPEND ON US ENTIRELY.
BUT IF ANTS DISAPPEARED, TENS OF THOUSANDS OF OTHER PLANT AND ANIMAL SPECIES WOULD DIE OFF AS WELL, WEAKENING LAND ECOSYSTEMS ALMOST EVERYWHERE.

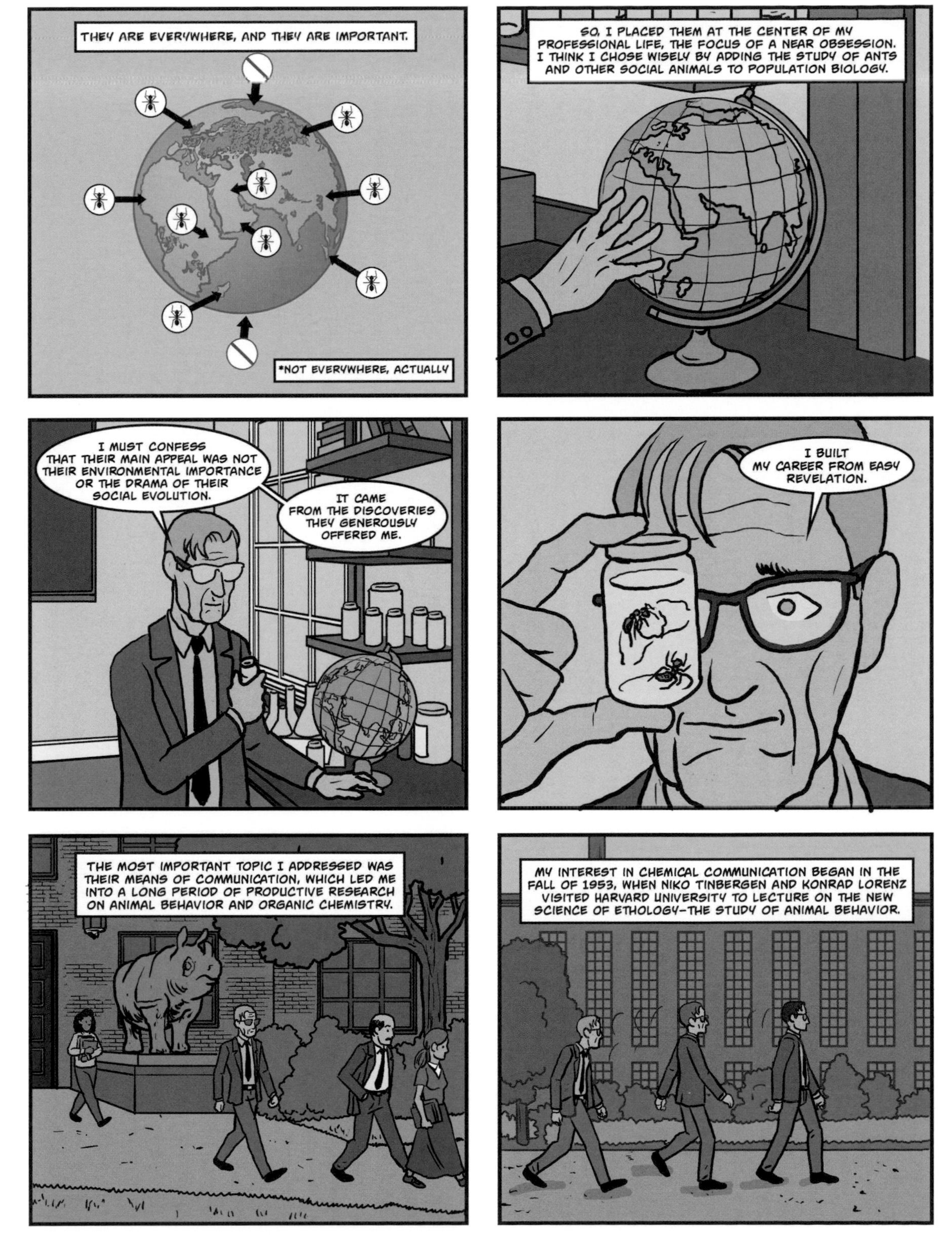
THEY ARE EVERYWHERE, AND THEY ARE IMPORTANT.
*NOT EVERYWHERE, ACTUALLY
SO, I PLACED THEM AT THE CENTER OF MY PROFESSIONAL LIFE, THE FOCUS OF A NEAR OBSESSION. I THINK I CHOSE WISELY BY ADDING THE STUDY OF ANTS AND OTHER SOCIAL ANIMALS TO POPULATION BIOLOGY.
I MUST CONFESS THAT THEIR MAIN APPEAL WAS NOT THEIR ENVIRONMENTAL IMPORTANCE OR THE DRAMA OF THEIR SOCIAL EVOLUTION.
IT CAME FROM THE DISCOVERIES THEY GENEROUSLY OFFERED ME.
I BUILT MY CAREER FROM EASY REVELATION.
THE MOST IMPORTANT TOPIC I ADDRESSED WAS THEIR MEANS OF COMMUNICATION, WHICH LED ME INTO A LONG PERIOD OF PRODUCTIVE RESEARCH ON ANIMAL BEHAVIOR AND ORGANIC CHEMISTRY.
MY INTEREST IN CHEMICAL COMMUNICATION BEGAN IN THE FALL OF 1953, WHEN NIKO TINBERGEN AND KONRAD LORENZ VISITED HARVARD UNIVERSITY TO LECTURE ON THE NEW SCIENCE OF ETHOLOGY—THE STUDY OF ANIMAL BEHAVIOR.

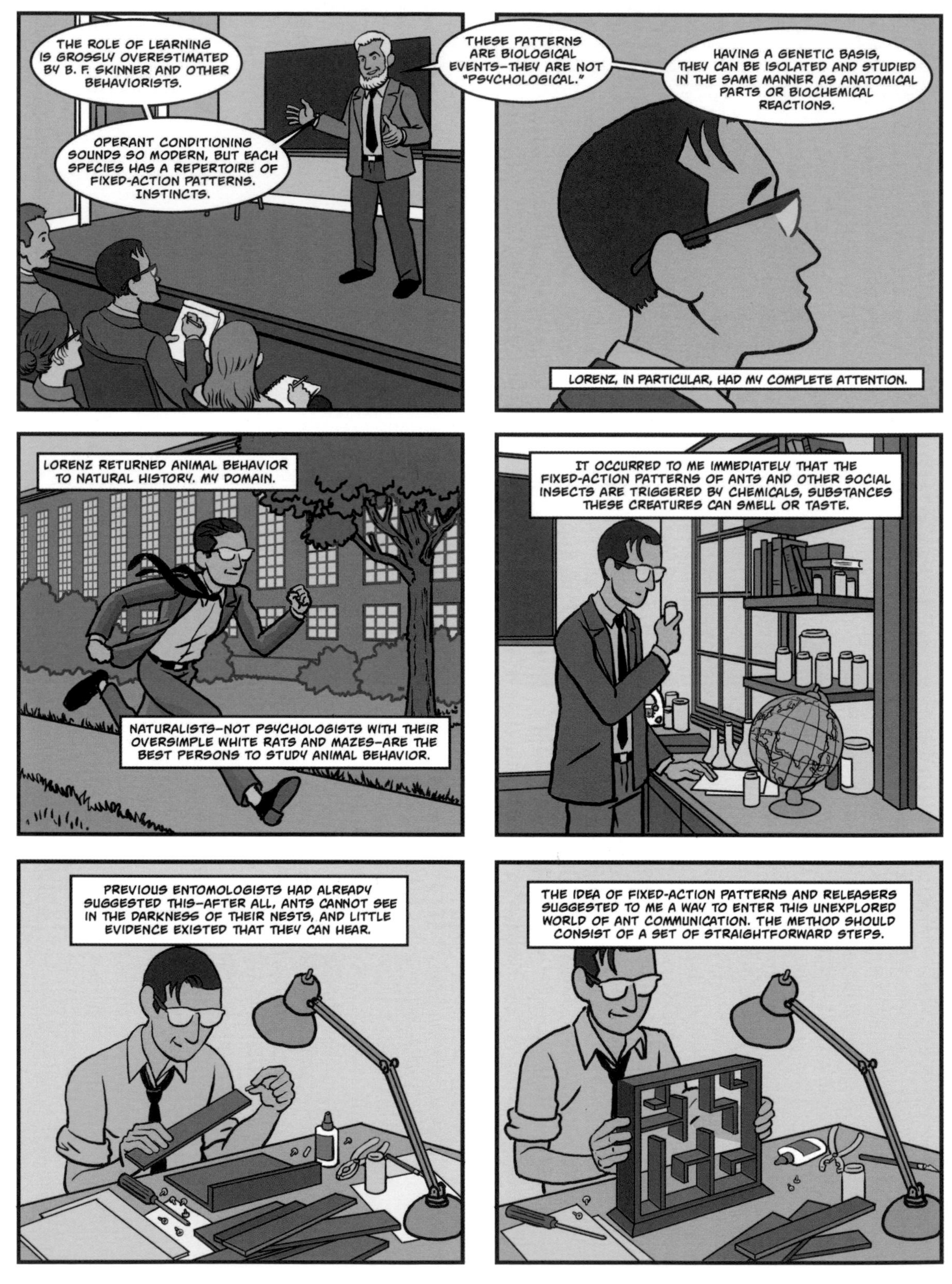
THE ROLE OF LEARNING IS GROSSLY OVERESTIMATED BY B. F. SKINNER AND OTHER BEHAVIORISTS.
OPERANT CONDITIONING SOUNDS SO MODERN, BUT EACH SPECIES HAS A REPERTOIRE OF FIXED-ACTION PATTERNS. INSTINCTS.
THESE PATTERNS ARE BIOLOGICAL EVENTS—THEY ARE NOT "PSYCHOLOGICAL."
HAVING A GENETIC BASIS, THEY CAN BE ISOLATED AND STUDIED IN THE SAME MANNER AS ANATOMICAL PARTS OR BIOCHEMICAL REACTIONS.
LORENZ, IN PARTICULAR, HAD MY COMPLETE ATTENTION.
LORENZ RETURNED ANIMAL BEHAVIOR TO NATURAL HISTORY. MY DOMAIN.
NATURALISTS—NOT PSYCHOLOGISTS WITH THEIR OVERSIMPLE WHITE RATS AND MAZES—ARE THE BEST PERSONS TO STUDY ANIMAL BEHAVIOR.
IT OCCURRED TO ME IMMEDIATELY THAT THE FIXED-ACTION PATTERNS OF ANTS AND OTHER SOCIAL INSECTS ARE TRIGGERED BY CHEMICALS, SUBSTANCES THESE CREATURES CAN SMELL OR TASTE.
PREVIOUS ENTOMOLOGISTS HAD ALREADY SUGGESTED THIS—AFTER ALL, ANTS CANNOT SEE IN THE DARKNESS OF THEIR NESTS, AND LITTLE EVIDENCE EXISTED THAT THEY CAN HEAR.
THE IDEA OF FIXED-ACTION PATTERNS AND RELEASERS SUGGESTED TO ME A WAY TO ENTER THIS UNEXPLORED WORLD OF ANT COMMUNICATION. THE METHOD SHOULD CONSIST OF A SET OF STRAIGHTFORWARD STEPS.

BREAK ANT SOCIAL BEHAVIOR INTO FIXED-ACTION PATTERNS, THEN, BY TRIAL AND ERROR, DETERMINE WHICH SECRETIONS CONTAIN THE TRIGGERING STIMULI.
PATTERN
STIMULI
SECRETION
FINALLY, SEPARATE AND IDENTIFY THE ACTIVE CHEMICALS IN THE SECRETIONS.

AS FAR AS I KNEW, I WAS THE ONLY PERSON THINKING ALONG THESE LINES. IN CAMBRIDGE, WITH A WELL-EQUIPPED LABORATORY, I BEGAN THE SEARCH FOR THE CHEMICAL RELEASERS OF ANT COMMUNICATION.

I STARTED WITH MY FAVORITE SPECIES, THE IMPORTED FIRE ANT.

THE MOST CONSPICUOUS FORM OF COMMUNICATION IN FIRE ANTS IS THE LAYING OF ODOR TRAILS TO FOOD.

WHEN THEY ENCOUNTER A PARTICLE OF FOOD TOO BIG OR AWKWARD TO CARRY HOME IN ONE TRIP, THEY HEAD BACK TO THE NEST IN A MORE OR LESS DIRECT LINE WHILE LAYING AN ODOR TRAIL.

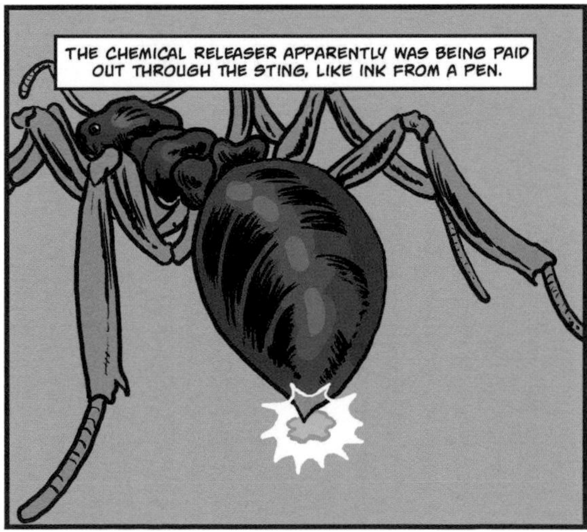
THE CHEMICAL RELEASER APPARENTLY WAS BEING PAID OUT THROUGH THE STING, LIKE INK FROM A PEN.

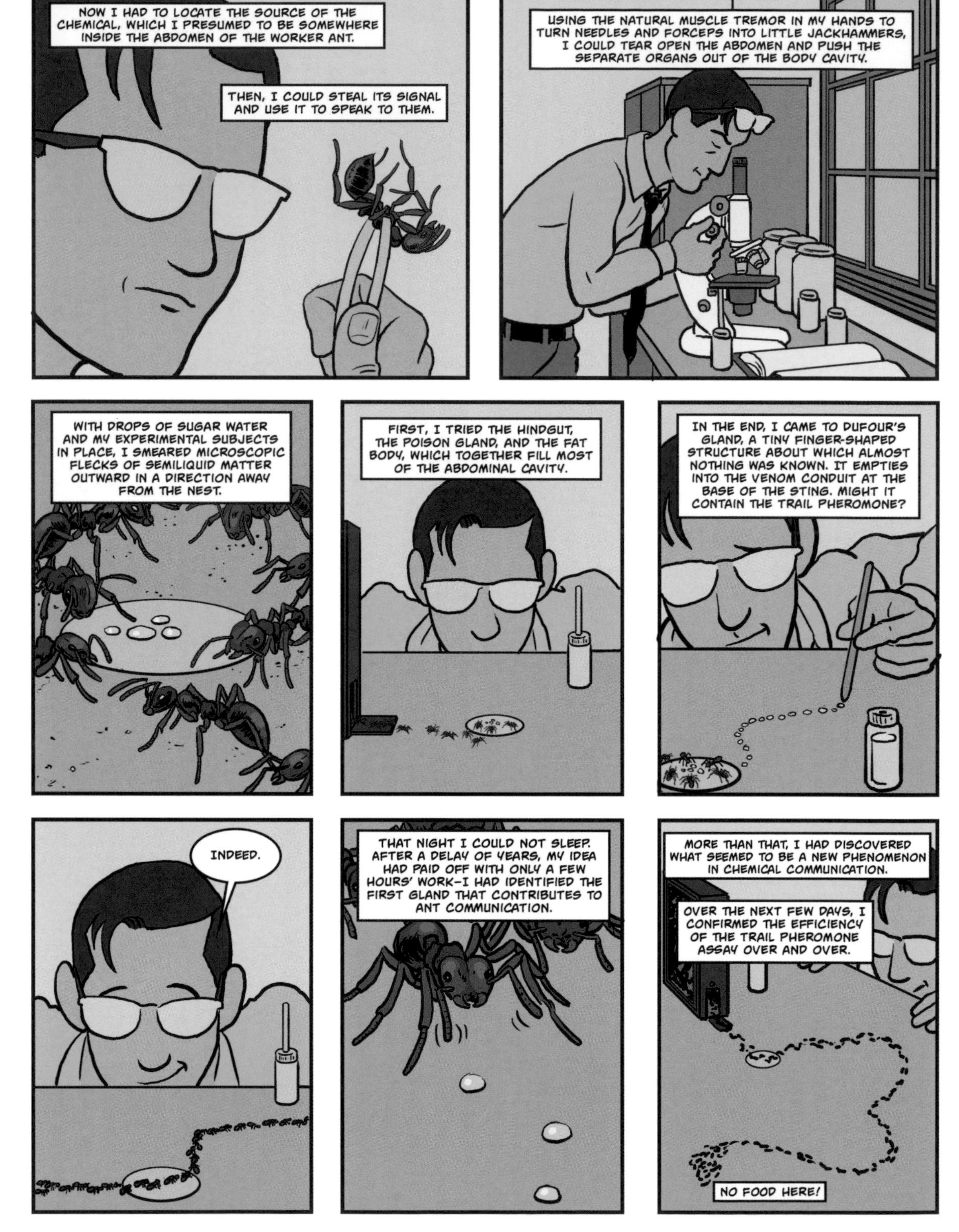
NOW I HAD TO LOCATE THE SOURCE OF THE CHEMICAL, WHICH I PRESUMED TO BE SOMEWHERE INSIDE THE ABDOMEN OF THE WORKER ANT.
THEN, I COULD STEAL ITS SIGNAL AND USE IT TO SPEAK TO THEM.
USING THE NATURAL MUSCLE TREMOR IN MY HANDS TO TURN NEEDLES AND FORCEPS INTO LITTLE JACKHAMMERS, I COULD TEAR OPEN THE ABDOMEN AND PUSH THE SEPARATE ORGANS OUT OF THE BODY CAVITY.
WITH DROPS OF SUGAR WATER AND MY EXPERIMENTAL SUBJECTS IN PLACE, I SMEARED MICROSCOPIC FLECKS OF SEMILIQUID MATTER OUTWARD IN A DIRECTION AWAY FROM THE NEST.
FIRST, I TRIED THE HINDGUT, THE POISON GLAND, AND THE FAT BODY, WHICH TOGETHER FILL MOST OF THE ABDOMINAL CAVITY.
IN THE END, I CAME TO DUFOUR'S GLAND, A TINY FINGER-SHAPED STRUCTURE ABOUT WHICH ALMOST NOTHING WAS KNOWN. IT EMPTIES INTO THE VENOM CONDUIT AT THE BASE OF THE STING. MIGHT IT CONTAIN THE TRAIL PHEROMONE?
INDEED.
THAT NIGHT I COULD NOT SLEEP. AFTER A DELAY OF YEARS, MY IDEA HAD PAID OFF WITH ONLY A FEW HOURS' WORK—I HAD IDENTIFIED THE FIRST GLAND THAT CONTRIBUTES TO ANT COMMUNICATION.
MORE THAN THAT, I HAD DISCOVERED WHAT SEEMED TO BE A NEW PHENOMENON IN CHEMICAL COMMUNICATION.
OVER THE NEXT FEW DAYS, I CONFIRMED THE EFFICIENCY OF THE TRAIL PHEROMONE ASSAY OVER AND OVER.
NO FOOD HERE!

IN SCIENCE, THERE IS NOTHING MORE PLEASANT THAN REPEATING AN EXPERIMENT THAT WORKS.
WHEN I LET A CONCENTRATED VAPOR MADE FROM MANY ANTS WAFT DOWN ONTO THE NEST, A LARGE PERCENTAGE OF THE WORKER FORCE EMERGED AND SPREAD OUT IN APPARENT SEARCH OF FOOD.
NOW, CHEMICAL ANALYSIS. FOR THAT I NEEDED TENS OF THOUSANDS OR HUNDREDS OF THOUSANDS OF ANTS, EACH WITH ITS VANISHING TRACE OF PHEROMONE.
BASED ON MY FIELD EXPERIENCE, I KNEW OF AN EASY WAY TO GET ENOUGH.
WHEN FIRE ANT NESTS ARE FLOODED IN NATURE BY RISING STREAM WATERS, THE WORKERS FLOAT TO THE SURFACE IN TIGHTLY PACKED MASSES. THEIR BODIES FORM A LIVING RAFT WITHIN WHICH THE QUEEN AND BROOD ARE SAFELY TUCKED.

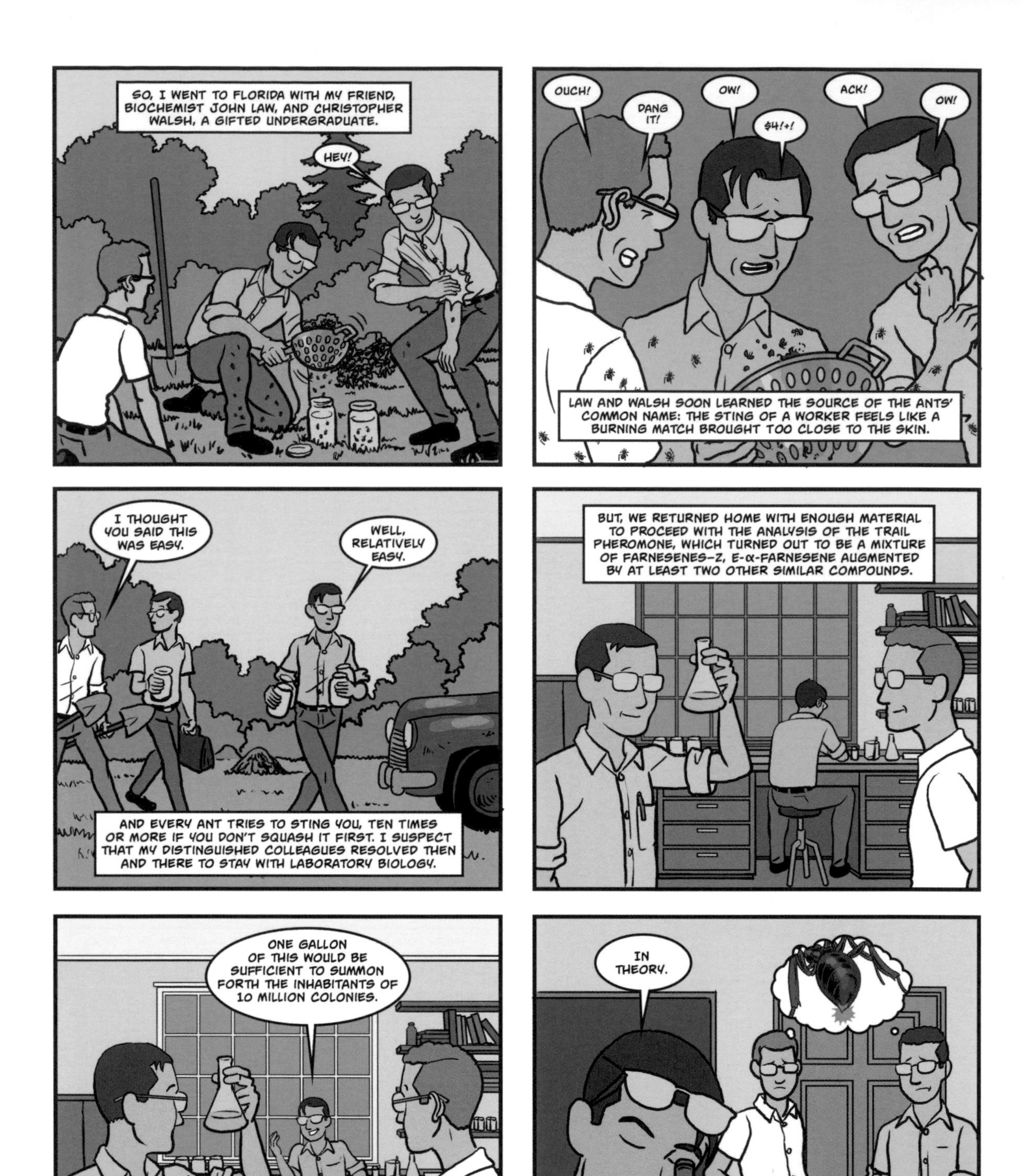
SO, I WENT TO FLORIDA WITH MY FRIEND, BIOCHEMIST JOHN LAW, AND CHRISTOPHER WALSH, A GIFTED UNDERGRADUATE.
HEY!
OUCH!
DANG IT!
OW!
$4!+!
ACK!
OW!
LAW AND WALSH SOON LEARNED THE SOURCE OF THE ANTS' COMMON NAME: THE STING OF A WORKER FEELS LIKE A BURNING MATCH BROUGHT TOO CLOSE TO THE SKIN.
I THOUGHT YOU SAID THIS WAS EASY.
WELL, RELATIVELY EASY.
AND EVERY ANT TRIES TO STING YOU, TEN TIMES OR MORE IF YOU DON'T SQUASH IT FIRST. I SUSPECT THAT MY DISTINGUISHED COLLEAGUES RESOLVED THEN AND THERE TO STAY WITH LABORATORY BIOLOGY.
BUT, WE RETURNED HOME WITH ENOUGH MATERIAL TO PROCEED WITH THE ANALYSIS OF THE TRAIL PHEROMONE, WHICH TURNED OUT TO BE A MIXTURE OF FARNESENES–Z, E-α-FARNESENE AUGMENTED BY AT LEAST TWO OTHER SIMILAR COMPOUNDS.
ONE GALLON OF THIS WOULD BE SUFFICIENT TO SUMMON FORTH THE INHABITANTS OF 10 MILLION COLONIES.
IN THEORY.

AS I LOOKED MORE CLOSELY AT THE FIRE ANT TRAIL, I STUMBLED ON A SECOND PHENOMENON OF SOCIAL BEHAVIOR—MASS COMMUNICATION.
THE AMOUNT OF FOOD OR THE SIZE OF AN ENEMY FORCE CANNOT BE TRANSMITTED BY SIGNALS FROM A SINGLE SCOUT.
THIS CAN BE CONVEYED ONLY BY GROUPS OF WORKERS SIGNALING TO OTHER GROUPS BY LAYING TRAILS ON TOP OF ONE ANOTHER OVER A SHORT INTERVAL OF TIME.
THE INFORMATION CONTAINED IN THE COMBINED ACTION OF MASSES OF INDIVIDUALS COMING AND GOING TO A TARGET IS SURPRISINGLY PRECISE.
LATER WRITERS POINTED TO A PARALLEL ACTION IN MASSES OF BRAIN CELLS, AND THE SIMILARITY THAT EXISTS BETWEEN THE BRAIN, THE ORGAN OF THOUGHT, AND THE INSECT COLONY, THE SUPERORGANISM.
THE QUESTION THEN AROSE, AND HAS SINCE BEEN ASKED MANY TIMES...
DOES THE RESEMBLANCE MEAN THAT AN ANT COLONY CAN SOMEHOW "THINK"?
I THINK NOT. THERE ARE TOO FEW ANTS, AND THOSE ARE TOO LOOSELY ORGANIZED TO FORM A BRAIN.

THE MOST BIZARRE PHEROMONE, IF THE GENERIC TERM CAN EVEN BE USED IN THIS CASE, IS THE SIGNAL OF THE DEAD—THE MEANS BY WHICH A CORPSE "ANNOUNCES" ITS NEW STATUS TO NESTMATES.
WHEN AN ANT DIES, AND IF IT HAS NOT BEEN CRUSHED OR TORN APART, IT SIMPLY CRUMPLES UP AND LIES STILL. ALTHOUGH ITS POSTURE AND INACTIVITY ARE ABNORMAL, NESTMATES CONTINUE TO WALK BY IT AS THOUGH NOTHING HAS HAPPENED.
TWO OR THREE DAYS PASS BEFORE RECOGNITION DAWNS, AND THEN IT IS THROUGH THE SMELL OF DECOMPOSITION.
OLEIC ACID
O
OH
MAYBE, WITH THE RIGHT CHEMICALS, I COULD CREATE AN ARTIFICIAL CORPSE.
AND SO FOR WEEKS, MY LABORATORY SMELLED LIKE A SEWER, GARBAGE DUMP, AND LOCKER ROOM. COMBINED.
IN CONTRAST TO THE RESPONSES OF MY NOSE—AND BRAIN—THE ANTS' RESPONSES TO THE CHEMICALS WERE CONSISTENTLY NARROW. THEY REMOVED ONLY THE PAPER SCRAPS TREATED WITH OLEIC ACID.
ANTS ARE NEITHER AESTHETIC NOR METICULOUSLY CLEAN IN ANY HUMAN SENSE. THEY ONLY REACT TO CUES THAT RELIABLY IDENTIFY A DECAYING BODY.
SO...WHAT WOULD HAPPEN IF A CORPSE CAME TO LIFE?

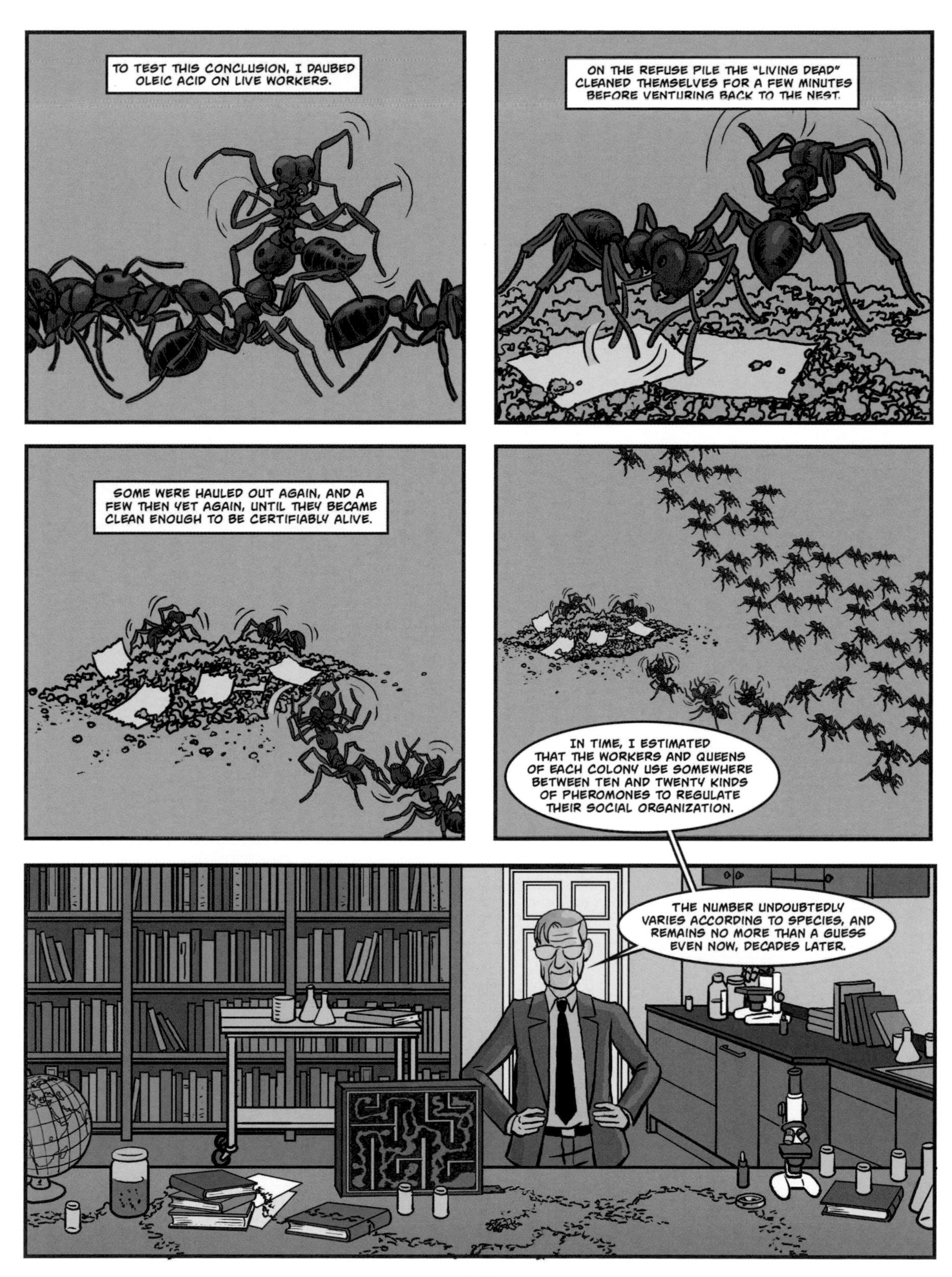
TO TEST THIS CONCLUSION, I DAUBED OLEIC ACID ON LIVE WORKERS.
ON THE REFUSE PILE THE "LIVING DEAD" CLEANED THEMSELVES FOR A FEW MINUTES BEFORE VENTURING BACK TO THE NEST.
SOME WERE HAULED OUT AGAIN, AND A FEW THEN YET AGAIN, UNTIL THEY BECAME CLEAN ENOUGH TO BE CERTIFIABLY ALIVE.
IN TIME, I ESTIMATED THAT THE WORKERS AND QUEENS OF EACH COLONY USE SOMEWHERE BETWEEN TEN AND TWENTY KINDS OF PHEROMONES TO REGULATE THEIR SOCIAL ORGANIZATION.
THE NUMBER UNDOUBTEDLY VARIES ACCORDING TO SPECIES, AND REMAINS NO MORE THAN A GUESS EVEN NOW, DECADES LATER.

NOW WE HAVE COME TO 1969. FOR SOME, IT IS THE EASILY REMEMBERED YEAR WHEN THE PHARAOH'S ANTS BEGAN TO STEAL CULTURE MEDIA FROM THE MOLECULAR BIOLOGISTS...
FOR ME, IT WAS THE YEAR THAT A YOUNG SCIENTIST WITH WHOM I HAD BEEN CORRESPONDING KNOCKED AT THE DOOR OF MY OFFICE IN THE BIOLOGICAL LABORATORIES.
BERT HÖLLDOBLER? ED WILSON.
A PLEASURE.
ZOOLOGIST, UNIVERSITY OF FRANKFURT
I WAS ABOUT TO ENTER THE MOST SUSTAINED AND PRODUCTIVE COLLABORATION OF MY RESEARCH CAREER, BUILT UPON A CLOSE FRIENDSHIP AND A COMMON LIFELONG COMMITMENT TO STUDYING ANTS.
WE REPRESENTED TWO CULTURES IN BEHAVIORAL BIOLOGY. THE STUDY OF ANIMAL BEHAVIOR—ETHOLOGY—WAS BERT'S FORTE. POPULATION BIOLOGY, WITH ITS RADICALLY DIFFERENT APPROACH TO BEHAVIOR, WAS BECOMING MINE.
HOME OF ETHOLOGY
HOME OF POPULATION BIOLOGY
AND AN INSECT COLONY IS A POPULATION. SOME COLONIES, LIKE THE QUEEN AND 20-MILLION-WORKER FORCE OF THE AFRICAN DRIVER ANT, HAVE MORE INHABITANTS THAN ENTIRE COUNTRIES.
LIKE HUMAN POPULATIONS, THE ONLY WAY TO UNDERSTAND SUCH ENSEMBLES FULLY IS TO TRACE THE LIVES AND DEATHS OF THEIR SEPARATE MEMBERS.

THE FINAL ELEMENT IN THE MIX IS EVOLUTION, YES? THE BEHAVIORS AND POPULATION ANALYSES ARE THE HISTORICAL PRODUCTS OF NATURAL SELECTION.
PUT TOGETHER, POPULATION BIOLOGY, ETHOLOGY, AND EVOLUTIONARY THEORY FORM THE CONTENT OF A NEW DISCIPLINE.
SOCIOBIOLOGY, WHICH I HAD DEFINED BACK IN 1971, IS THE SYSTEMATIC STUDY OF THE BIOLOGICAL BASIS OF SOCIAL BEHAVIOR AND OF THE ORGANIZATION OF COMPLEX SOCIETIES.
END THE WAR
END VIETNAM
AS OUR WORK TOGETHER PROGRESSED, BERT HÖLLDOBLER AND I EDGED TOWARD SOCIOBIOLOGY.
WE WERE, HOWEVER, FIRST AND FOREMOST ENTOMOLOGISTS.
WE SHARED THE CONVICTION THAT ANTS ARE WORTHY OF STUDY NO MATTER HOW IT IS DONE–BY ETHOLOGY, SOCIOBIOLOGY, OR ANY OTHER BIOLOGICAL DISCIPLINE.
BERT, ABOVE ALL, WAS ROOTED TO THE EARTH.
A NATURALIST.
HE WAS–AND STILL IS–THE MOST HONEST SCIENTIST I HAVE EVER KNOWN.
OKAY, THAT'S –AH–SIXTY-THREE SECONDS, THE FORAGER JUST ENTERED THE NEST, GOT IT?
YES. NOW, I WANT TO GO BACK AND SAY JUST ONE MORE THING ABOUT HENNIG AND THE ORIGINAL IDEA OF CLADISM...

AS WE FILLED THE HOURS AROUND TEDIOUS REPLICATE EXPERIMENTS...
OKAY, LET'S TIME THE NEXT ONE.
...HE ENDEAVORED TO MAKE EVERY DATUM IN HIS NOTEBOOK, EVERY NUANCE OF EXPRESSION IN HIS PUBLISHED REPORTS, AS STRAIGHT AND TRANSPARENT AS HE COULD.

WAIT, WHAT IF WE...
IN SCIENCE, OBSESSIVENESS CAN BE A VIRTUE...IF KEPT UNDER CONTROL.
TO A DEGREE I HAVE NOT ENCOUNTERED ELSEWHERE, HÖLLDOBLER EXTENDED THIS URGE TO THE DESIGN OF EXPERIMENTS AND WEIGHING OF EVIDENCE.

MANY SUCCESSFUL RESEARCHERS STOP WITH A SINGLE WELL-CONDUCTED PROCEDURE, WHICH THEY REPEAT OFTEN ENOUGH FOR THE OVERALL RESULT TO BE STATISTICALLY PERSUASIVE.
THAT'S PLENTY. I THINK IT LIKELY THAT SUCH AND SUCH IS THE CASE.

OTHERS HOLD BACK.
WHAT DIFFERENT EXPERIMENT CAN I PERFORM, USING NEW KINDS OF MEASUREMENTS, THAT WILL TEST THE CONCLUSION MORE RIGOROUSLY?

IF THEY THEN PERFORM THE SECOND EXPERIMENT AND FIND THE SECOND RESULT CONSISTENT WITH THE FIRST, THEN...
THAT PRETTY WELL PROVES IT. LET'S MOVE ON.

HÖLLDOBLER IS A MEMBER OF THE SECOND GROUP. BUT SOMETIMES HE WOULD PAUSE YET AGAIN...
IS THERE A THIRD WAY?
SERIOUSLY?
HE WAS THE ONLY THIRD-WAY RESEARCHER I HAVE EVER KNOWN.

AND IF NEW DATA DID NOT FIT? HE IS ONE OF THE FEW SCIENTISTS I HAVE KNOWN ACTUALLY WILLING TO ABANDON A HYPOTHESIS.
A SCIENTIST'S SCIENTIST.
IN 1985, WE MADE OUR FIRST FIELD TRIP TOGETHER TO COSTA RICA. WE DROVE NORTH FROM SAN JOSÉ TO LA SELVA, THE FIELD STATION OF THE ORGANIZATION FOR TROPICAL STUDIES.
AS WE ENTERED THE RAIN FOREST, I USED MY MORE GENERAL KNOWLEDGE OF ANTS TO FIND AND IDENTIFY COLONIES THAT MIGHT BE OF EXCEPTIONAL INTEREST IN BEHAVIORAL WORK.
I WAS LOOKING FOR A QUICK AND EXCITING PAYOFF.
ONE CANDIDATE WAS THE PRIMITIVE GENUS I FOUND NESTING IN ROTTING LOGS.
NO COLONIES HAD EVER PREVIOUSLY BEEN STUDIED IN LIFE. I WAS EAGER TO RECORD THE KEY FACTS OF THE SOCIAL BEHAVIOR OF THIS ANT.
WE PLUNGED INTO THE WORK, TAKING NOTES ON COLONY SIZE, THE NUMBER OF QUEENS, DIVISION OF LABOR, AND THE KINDS OF INSECTS AND OTHER SMALL ANIMALS CAPTURED BY THE WORKERS.
PRIONOPELTA
CAMPODEID DIPLURANS

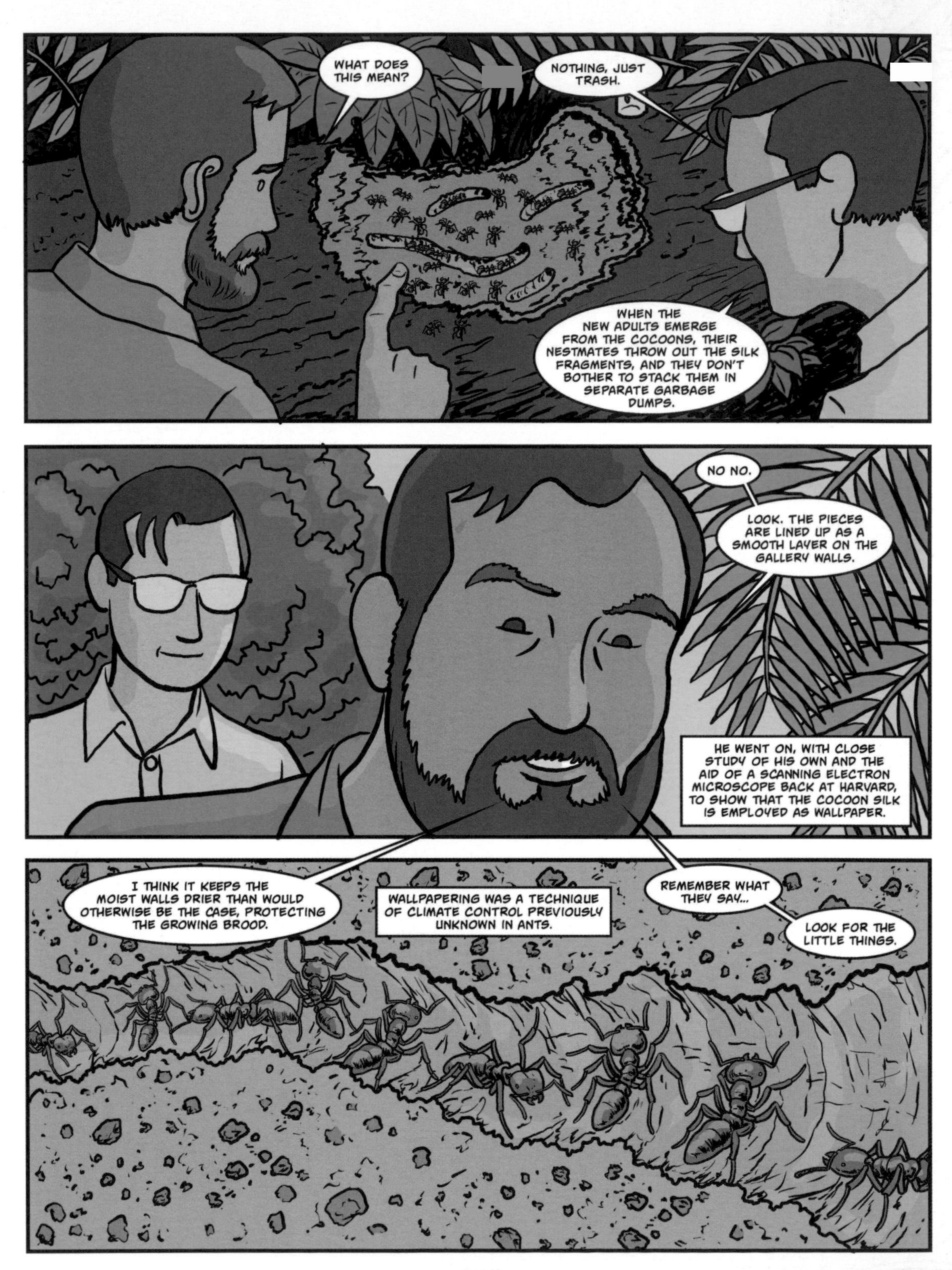
WHAT DOES THIS MEAN?
NOTHING, JUST TRASH.
WHEN THE NEW ADULTS EMERGE FROM THE COCOONS, THEIR NESTMATES THROW OUT THE SILK FRAGMENTS, AND THEY DON'T BOTHER TO STACK THEM IN SEPARATE GARBAGE DUMPS.
NO NO.
LOOK. THE PIECES ARE LINED UP AS A SMOOTH LAYER ON THE GALLERY WALLS.
HE WENT ON, WITH CLOSE STUDY OF HIS OWN AND THE AID OF A SCANNING ELECTRON MICROSCOPE BACK AT HARVARD, TO SHOW THAT THE COCOON SILK IS EMPLOYED AS WALLPAPER.
I THINK IT KEEPS THE MOIST WALLS DRIER THAN WOULD OTHERWISE BE THE CASE, PROTECTING THE GROWING BROOD.
WALLPAPERING WAS A TECHNIQUE OF CLIMATE CONTROL PREVIOUSLY UNKNOWN IN ANTS.
REMEMBER WHAT THEY SAY...
LOOK FOR THE LITTLE THINGS.

LOOK, SOME OF THE FORAGING WORKERS APPEAR TO BE MOVING MORE SLOWLY WHILE THEY DRAG THEIR HIND LEGS.
SO?
INDIVIDUAL ANTS OFTEN MOVE SLOWLY OR ERRATICALLY FOR NO PARTICULAR GOOD REASON.
AND REMEMBER, THESE *PRIONOPELTA* ARE PRIMITIVE. THERE'S NO REASON TO BELIEVE THEY LAY ODOR TRAILS IN ANY CASE.
BUT HÖLLDOBLER PERSEVERED. HE FOUND THAT NOT ONLY DO THE WORKERS LAY ODOR TRAILS—TO RECRUIT NESTMATES TO NEW NEST SITES—BUT THE ATTRACTIVE SUBSTANCE COMES FROM A PREVIOUSLY UNSUSPECTED GLAND LOCATED IN THE HIND LEGS.
THE PHEROMONES ARE SMEARED IN A LINE AS THE ANTS DRAG THEIR HIND LEGS OVER THE GROUND.
THE EXISTENCE OF THE GLAND PROVIDED AN IMPORTANT CLUE TO THE EVOLUTIONARY RELATIONSHIPS OF *PRIONOPELTA*.
FROM THESE TWO WEEKS OF DATA GATHERED IN THE LA SELVA FOREST, WE WROTE MANY SCIENTIFIC ARTICLES, IN WHOLE OR IN PART.
AND DURING OUR YEARS OF COLLABORATION, WE MADE UNCOUNTED OTHER DISCOVERIES—FIRST ONE TAKING THE LEAD, THEN THE OTHER—WHILE GRUBBING AROUND AND TALKING BACK AND FORTH.

OUR PARTNERSHIP IN MOST RESPECTS ENDED WHEN BERT, AFTER BEING APPROACHED BY SEVERAL INSTITUTIONS IN EUROPE, WAS OFFERED A PROFESSORSHIP AT WÜRZBURG.
HE HAD A DESIRE, ALMOST AN OBSESSION, TO GET INSIDE THE MUSCLES, GLANDS, AND BRAINS OF ANTS TO LEARN HOW THESE ORGANS MEDIATE SOCIAL BEHAVIOR AND ORGANIZATION.
I WANT TO UNDERSTAND A THOUSAND "LITTLE THINGS" TO MAKE A GREAT WHOLE.
ONE DAY, AS BERT GREW MORE SERIOUS ABOUT LEAVING, WE DECIDED TO WRITE A BOOK RECOUNTING EVERYTHING WE KNEW ABOUT ANTS.
WHILE WE'RE AT IT, WHY NOT TRY FOR A BOOK THAT HAS EVERYTHING EVERYBODY EVER KNEW ABOUT ANTS...
...THROUGHOUT HISTORY.
SURE, WHY NOT? WE'LL PROBABLY FALL SHORT, BUT HEY.
IT IS A WORTHY CONCEIT.
LONG STORY SHORT, WE WROTE THAT BOOK.
AND BY LONG, I MEAN 732 DOUBLE-COLUMNED PAGES, HUNDREDS OF FIGURES AND COLOR PLATES, AND A BIBLIOGRAPHY OF 3,000 ENTRIES.
KATHLEEN HORTON GATHERED MUCH OF THE RESEARCH AND TYPED AND EDITED ALL OF IT.
THE PRINTED BOOK FULFILLS MY CRITERION OF A MAGNUM OPUS—A BOOK WHICH WHEN DROPPED FROM A THREE-STORY BUILDING IS BIG ENOUGH TO KILL SOMEONE.
THE ANTS
I CAN ASSURE YOU, WE DID NOT DO THIS EXPERIMENT, EVEN ONCE.

THE ANTS CAME OUT IN 1990, AND LONG STORY SHORT—AGAIN—ONE YEAR LATER, IT WON THE PULITZER PRIZE IN GENERAL NONFICTION. WHEN I HEARD, I CALLED BERT, WHO WAS NOW IN BAVARIA.
HOW DOES IT FEEL TO WIN AMERICA'S MOST FAMOUS LITERARY AWARD?
UNIVERSITY OF WÜRZBURG

YOU MEAN SCIENTIFIC, YES?
HIS ACCENT, WHICH HE NEVER LOST IN ALL HIS YEARS HERE IN AMERICA, WAS STRENGTHENING.
NO, NOT SCIENTIFIC.
LITERARY.

THUMP!

HOW DO YOU THINK IT FEELS?
WONDERFUL.
WONDERFUL.

IT WAS.
WHERE COULD I GO FROM HERE BUT DOWN?

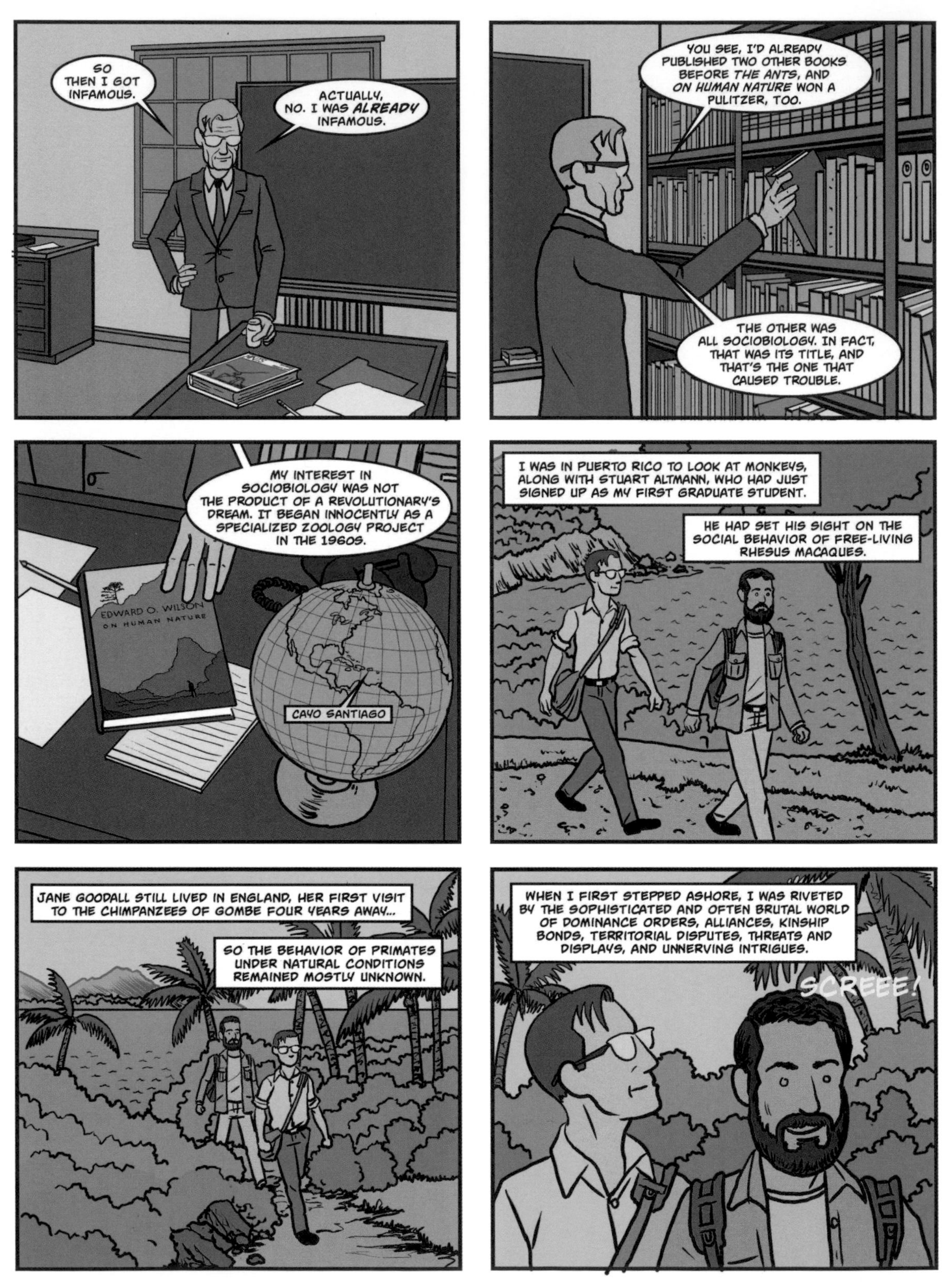
SO THEN I GOT INFAMOUS.
ACTUALLY, NO. I WAS ALREADY INFAMOUS.
YOU SEE, I'D ALREADY PUBLISHED TWO OTHER BOOKS BEFORE THE ANTS, AND ON HUMAN NATURE WON A PULITZER, TOO.
THE OTHER WAS ALL SOCIOBIOLOGY. IN FACT, THAT WAS ITS TITLE, AND THAT'S THE ONE THAT CAUSED TROUBLE.
MY INTEREST IN SOCIOBIOLOGY WAS NOT THE PRODUCT OF A REVOLUTIONARY'S DREAM. IT BEGAN INNOCENTLY AS A SPECIALIZED ZOOLOGY PROJECT IN THE 1960S.
EDWARD O. WILSON
ON HUMAN NATURE
CAYO SANTIAGO
I WAS IN PUERTO RICO TO LOOK AT MONKEYS, ALONG WITH STUART ALTMANN, WHO HAD JUST SIGNED UP AS MY FIRST GRADUATE STUDENT.
HE HAD SET HIS SIGHT ON THE SOCIAL BEHAVIOR OF FREE-LIVING RHESUS MACAQUES.
JANE GOODALL STILL LIVED IN ENGLAND, HER FIRST VISIT TO THE CHIMPANZEES OF GOMBE FOUR YEARS AWAY...
SO THE BEHAVIOR OF PRIMATES UNDER NATURAL CONDITIONS REMAINED MOSTLY UNKNOWN.
WHEN I FIRST STEPPED ASHORE, I WAS RIVETED BY THE SOPHISTICATED AND OFTEN BRUTAL WORLD OF DOMINANCE ORDERS, ALLIANCES, KINSHIP BONDS, TERRITORIAL DISPUTES, THREATS AND DISPLAYS, AND UNNERVING INTRIGUES.
SCREEE!

I KNEW ALMOST NOTHING ABOUT MACAQUE SOCIETIES.
TWO THINGS. DON'T MOVE TOO SUDDENLY NEAR AN INFANT, AS THOUGH YOU MEAN TO HARM IT. YOU MIGHT BE ATTACKED BY A MALE, AND SOME ARE BIG AND TOUGH.

AND IF YOU DO HAPPEN TO BE THREATENED, DON'T LOOK THE MALE IN THE FACE. A STARE IS A THREAT AND MIGHT PROVOKE AN ATTACK.
JUST HANG YOUR HEAD DOWN AND LOOK AWAY.

SURE ENOUGH, IN A CARELESS MOMENT ON THE SECOND DAY...

I FROZE, GENUINELY AFRAID. UNTIL NOW, I HAD THOUGHT OF MACAQUES AS HARMLESS LITTLE MONKEYS.
THIS ONE LOOKED, FOR THE MOMENT, LIKE A SMALL GORILLA.

SORRY, DIDN'T MEAN ANYTHING, SORRY.
AFTER A FEW MINUTES MY CHALLENGER LEFT.

A GENERAL THEORY, WE AGREED, MIGHT TAKE FORM UNDER THE NAME OF "SOCIOBIOLOGY." STUART WAS ALREADY USING THAT WORD TO DESCRIBE HIS STUDIES.
IN THE 1950S, THERE EXISTED NO THEORY TO EXPLAIN DIVERSITY–WHY VARIOUS BEHAVIORS HAVE ARISEN IN SOME GROUPS AND NOT OTHERS.
ALTMANN HAD A GOOD IDEA. HE INTENDED TO DEVISE PROBABILITY TRANSITION MATRICES OF BEHAVIORAL ACTS TO QUANTIFY SOCIAL INTERACTIONS.
Aggressive
Contested status
Neutral
Neutralized dominance
Submissive
AN IMPORTANT STEP, BUT WHERE DOES IT LEAD?
IT STILL OFFERS NO EXPLANATION OF HOW OR WHY A PARTICULAR SPECIES OF MONKEY OR ANT ARRIVED AT ONE PATTERN IN THE COURSE OF ITS EVOLUTION AS OPPOSED TO ANOTHER.
BY THE EARLY 1960S, I BEGAN TO SEE PROMISE IN POPULATION BIOLOGY AS A POSSIBLE FOUNDATION DISCIPLINE FOR SOCIOBIOLOGY.
AN ANIMAL SOCIETY IS A POPULATION, AND IT SHOULD BE POSSIBLE TO ANALYZE ITS STRUCTURE AND EVOLUTION AS PART OF POPULATION BIOLOGY.
I HAD ENTERED POPULATION BIOLOGY NOT TO SERVE SOCIOBIOLOGY BUT TO HELP FASHION A COUNTERWEIGHT TO MOLECULAR BIOLOGY.
POPULATIONS FOLLOW AT LEAST ***SOME*** LAWS DIFFERENT FROM THOSE OPERATING AT THE MOLECULAR LEVEL, LAWS THAT CANNOT BE CONSTRUCTED BY ANY LOGICAL PROGRESSION ***UPWARD*** FROM MOLECULAR BIOLOGY.

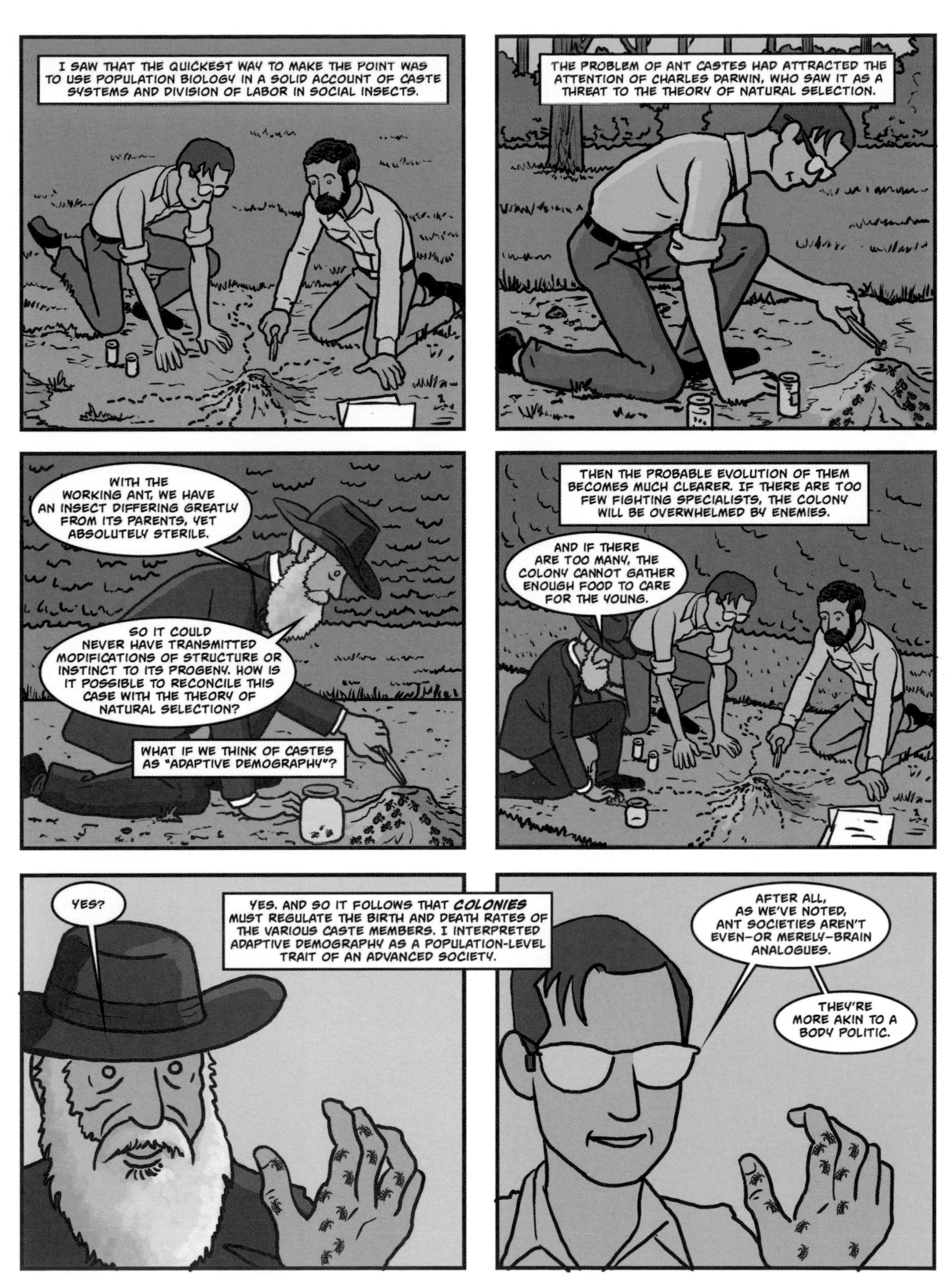
I SAW THAT THE QUICKEST WAY TO MAKE THE POINT WAS TO USE POPULATION BIOLOGY IN A SOLID ACCOUNT OF CASTE SYSTEMS AND DIVISION OF LABOR IN SOCIAL INSECTS.
THE PROBLEM OF ANT CASTES HAD ATTRACTED THE ATTENTION OF CHARLES DARWIN, WHO SAW IT AS A THREAT TO THE THEORY OF NATURAL SELECTION.
WITH THE WORKING ANT, WE HAVE AN INSECT DIFFERING GREATLY FROM ITS PARENTS, YET ABSOLUTELY STERILE.
SO IT COULD NEVER HAVE TRANSMITTED MODIFICATIONS OF STRUCTURE OR INSTINCT TO ITS PROGENY. HOW IS IT POSSIBLE TO RECONCILE THIS CASE WITH THE THEORY OF NATURAL SELECTION?
WHAT IF WE THINK OF CASTES AS "ADAPTIVE DEMOGRAPHY"?
THEN THE PROBABLE EVOLUTION OF THEM BECOMES MUCH CLEARER. IF THERE ARE TOO FEW FIGHTING SPECIALISTS, THE COLONY WILL BE OVERWHELMED BY ENEMIES.
AND IF THERE ARE TOO MANY, THE COLONY CANNOT GATHER ENOUGH FOOD TO CARE FOR THE YOUNG.
YES?
YES. AND SO IT FOLLOWS THAT *COLONIES* MUST REGULATE THE BIRTH AND DEATH RATES OF THE VARIOUS CASTE MEMBERS. I INTERPRETED ADAPTIVE DEMOGRAPHY AS A POPULATION-LEVEL TRAIT OF AN ADVANCED SOCIETY.
AFTER ALL, AS WE'VE NOTED, ANT SOCIETIES AREN'T EVEN—OR MERELY—BRAIN ANALOGUES.
THEY'RE MORE AKIN TO A BODY POLITIC.

IN 1968, AND WITH THE AID OF MODELS IN LINEAR PROGRAMMING, I REFINED THE IDEA OF ADAPTIVE DEMOGRAPHY AND DEVELOPED SEVERAL NEW PRINCIPLES OF CASTE EVOLUTION.
WE WERE ON OUR WAY TOWARD ATTAINING SOCIOBIOLOGY.

THE ELEMENTS CAME FROM MANY SOURCES, BUT WHEN THE MOST IMPORTANT IDEA OF ALL CAME ALONG, I, AT FIRST, RESISTED IT WITH ALL MY ABILITY.

I FIRST READ WILLIAM HAMILTON'S TWO-PART ARTICLE ON KIN SELECTION—"THE GENETICAL EVOLUTION OF SOCIAL BEHAVIOR"—DURING A TRIP FROM BOSTON TO MIAMI IN THE SPRING OF 1965.

I PULLED HAMILTON'S PAPER—WRITTEN WHILE HE WAS A STUDENT—OUT OF MY BRIEFCASE SOMEWHERE NORTH OF NEW HAVEN AND RIFFLED THROUGH IT IMPATIENTLY.

THE PROSE WAS CONVOLUTED AND THE FULL-DRESS MATHEMATICAL TREATMENT DIFFICULT, BUT I UNDERSTOOD HIS MAIN POINT QUICKLY ENOUGH.

CONVENTIONAL DARWINISM ENVISAGES NATURAL SELECTION AS AN EVENT OCCURRING DIRECTLY BETWEEN GENERATIONS, FROM PARENT TO OFFSPRING.

BECAUSE LINEAGES THAT SURVIVE AND REPRODUCE BETTER CREATE MORE OFFSPRING IN EACH GENERATION, THEIR GENES COMES TO PREDOMINATE IN THE POPULATION.
THIS IS—AGAIN, BY DEFINITION—EVOLUTION BY NATURAL SELECTION.

IN ONE IMPORTANT RESPECT, THIS TRADITIONAL PROCESS OF NATURAL SELECTION CAN BE CALLED A TYPE OF KIN SELECTION. PARENTS AND OFFSPRING ARE, AFTER ALL, CLOSE KIN.
BUT AS HAMILTON OBSERVED, BROTHERS, SISTERS, UNCLES, AUNTS, COUSINS, AND SO FORTH ARE ALSO KIN...
OBVIOUSLY.
...AND HE THOUGHT ABOUT WHAT THIS MEANS FOR EVOLUTION.
IF THERE IS ANY INTERACTION AMONG KIN INFLUENCED BY GENES—SAY, A HEREDITARY TENDENCY TOWARD ALTRUISM, OR COOPERATION, OR SIBLING RIVALRY—
THAT WILL RESULT IN A CHANGE IN SURVIVAL AND REPRODUCTION...
...AND SHOULD EQUALLY WELL CAUSE EVOLUTION BY NATURAL SELECTION.
PERHAPS THE ANCILLARY FORMS OF KIN SELECTION DRIVE MOST FORMS OF SOCIAL EVOLUTION?
WHAT MADE HAMILTON'S IDEA ATTRACTIVE WAS THAT IT HELPED TO RESOLVE THE CLASSIC PROBLEM OF SELF-SACRIFICE.
IT MIGHT SEEM ON FIRST THOUGHT THAT SELFISHNESS MUST REIGN IN THE LIVING WORLD, AND THAT COOPERATION CAN NEVER APPEAR EXCEPT TO ENHANCE SELFISH ENDS.
BUT IF AN ALTRUISTIC ACT HELPS RELATIVES, IT INCREASES THE SURVIVAL OF GENES THAT ARE IDENTICAL WITH THOSE OF THE ALTRUIST, JUST AS THE CASE IN PARENTS AND OFFSPRING.
THE BODY MAY DIE BECAUSE OF A SELFLESS ACTION, BUT THE GENES—INCLUDING THOSE THAT PRESCRIBE ALTRUISM—WILL FLOURISH.
THIS LEADS TO AN INTERESTING MENTAL GAME NOT UNLIKE CALCULATING THE ODDS IN GAMBLING. WHAT, FOR EXAMPLE, IS THE FRACTION OF GENES SHARED WITH A SECOND COUSIN, OR A HALF-SISTER'S FULL NIECE?
THIS NUMBER, THE DEGREE OF KINSHIP, HAMILTON SAW TO BE CRUCIAL IN THE EVOLUTION OF ALTRUISM.
YOU MIGHT RISK YOUR LIFE FOR A BROTHER, FOR EXAMPLE, BUT THE MOST YOU ARE LIKELY TO GIVE A THIRD COUSIN IS A PIECE OF ADVICE.

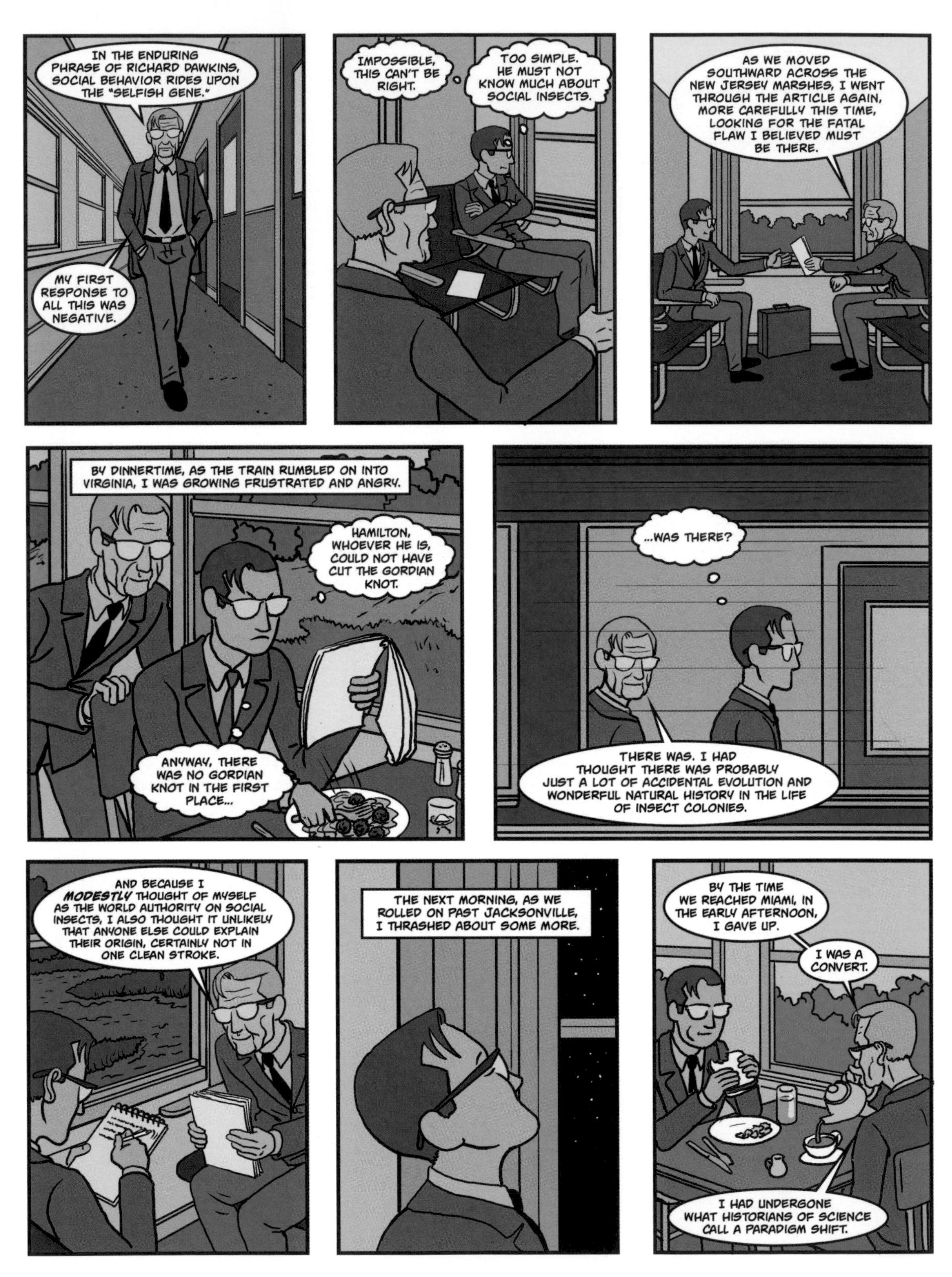
IN THE ENDURING PHRASE OF RICHARD DAWKINS, SOCIAL BEHAVIOR RIDES UPON THE "SELFISH GENE."
MY FIRST RESPONSE TO ALL THIS WAS NEGATIVE.
IMPOSSIBLE, THIS CAN'T BE RIGHT.
TOO SIMPLE. HE MUST NOT KNOW MUCH ABOUT SOCIAL INSECTS.
AS WE MOVED SOUTHWARD ACROSS THE NEW JERSEY MARSHES, I WENT THROUGH THE ARTICLE AGAIN, MORE CAREFULLY THIS TIME, LOOKING FOR THE FATAL FLAW I BELIEVED MUST BE THERE.
BY DINNERTIME, AS THE TRAIN RUMBLED ON INTO VIRGINIA, I WAS GROWING FRUSTRATED AND ANGRY.
HAMILTON, WHOEVER HE IS, COULD NOT HAVE CUT THE GORDIAN KNOT.
ANYWAY, THERE WAS NO GORDIAN KNOT IN THE FIRST PLACE...
...WAS THERE?
THERE WAS. I HAD THOUGHT THERE WAS PROBABLY JUST A LOT OF ACCIDENTAL EVOLUTION AND WONDERFUL NATURAL HISTORY IN THE LIFE OF INSECT COLONIES.
AND BECAUSE I *MODESTLY* THOUGHT OF MYSELF AS THE WORLD AUTHORITY ON SOCIAL INSECTS, I ALSO THOUGHT IT UNLIKELY THAT ANYONE ELSE COULD EXPLAIN THEIR ORIGIN, CERTAINLY NOT IN ONE CLEAN STROKE.
THE NEXT MORNING, AS WE ROLLED ON PAST JACKSONVILLE, I THRASHED ABOUT SOME MORE.
BY THE TIME WE REACHED MIAMI, IN THE EARLY AFTERNOON, I GAVE UP.
I WAS A CONVERT.
I HAD UNDERGONE WHAT HISTORIANS OF SCIENCE CALL A PARADIGM SHIFT.

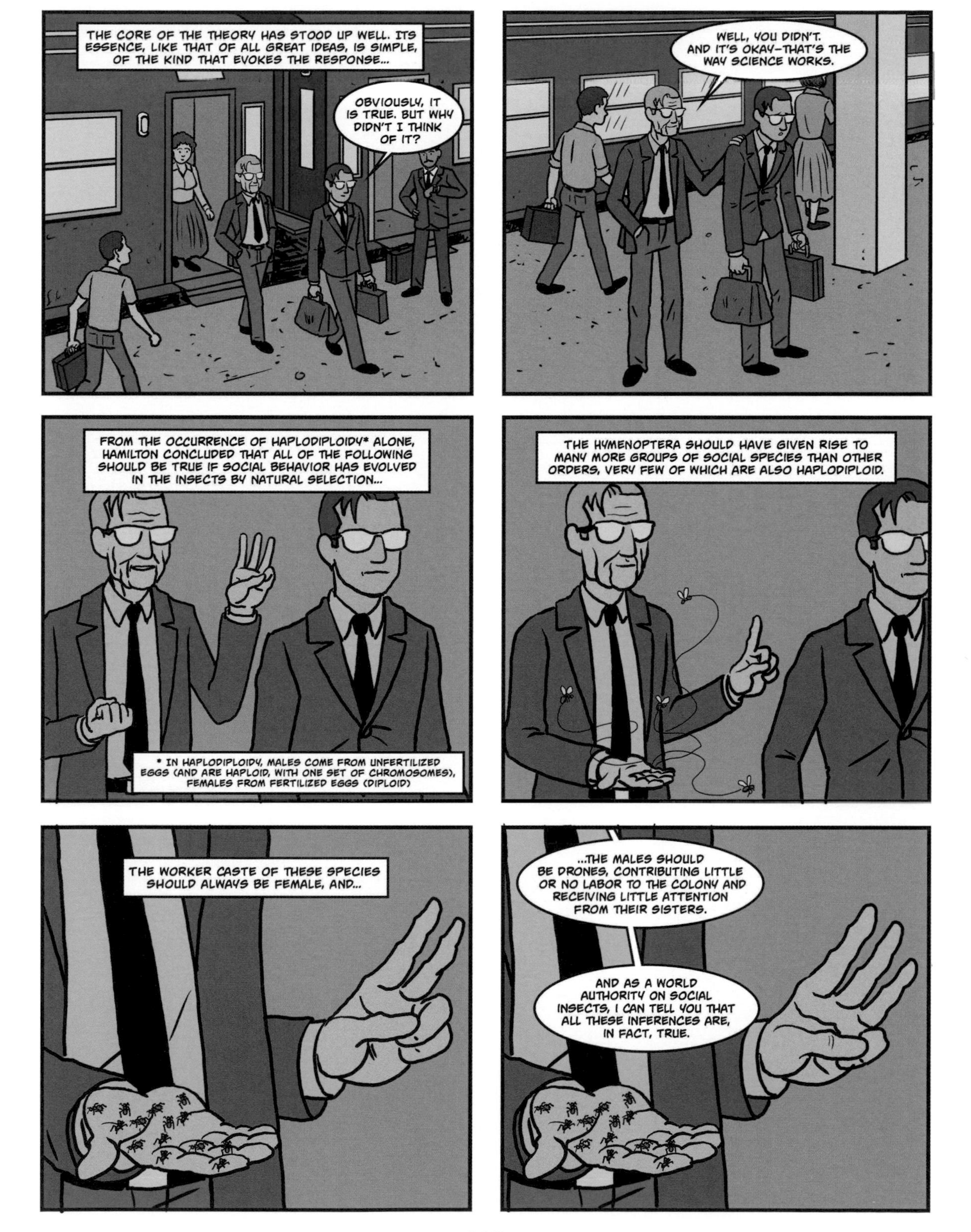
THE CORE OF THE THEORY HAS STOOD UP WELL. ITS ESSENCE, LIKE THAT OF ALL GREAT IDEAS, IS SIMPLE, OF THE KIND THAT EVOKES THE RESPONSE...
OBVIOUSLY, IT IS TRUE. BUT WHY DIDN'T I THINK OF IT?
WELL, YOU DIDN'T. AND IT'S OKAY—THAT'S THE WAY SCIENCE WORKS.
FROM THE OCCURRENCE OF HAPLODIPLOIDY* ALONE, HAMILTON CONCLUDED THAT ALL OF THE FOLLOWING SHOULD BE TRUE IF SOCIAL BEHAVIOR HAS EVOLVED IN THE INSECTS BY NATURAL SELECTION...
* IN HAPLODIPLOIDY, MALES COME FROM UNFERTILIZED EGGS (AND ARE HAPLOID, WITH ONE SET OF CHROMOSOMES), FEMALES FROM FERTILIZED EGGS (DIPLOID)
THE HYMENOPTERA SHOULD HAVE GIVEN RISE TO MANY MORE GROUPS OF SOCIAL SPECIES THAN OTHER ORDERS, VERY FEW OF WHICH ARE ALSO HAPLODIPLOID.
THE WORKER CASTE OF THESE SPECIES SHOULD ALWAYS BE FEMALE, AND...
...THE MALES SHOULD BE DRONES, CONTRIBUTING LITTLE OR NO LABOR TO THE COLONY AND RECEIVING LITTLE ATTENTION FROM THEIR SISTERS.
AND AS A WORLD AUTHORITY ON SOCIAL INSECTS, I CAN TELL YOU THAT ALL THESE INFERENCES ARE, IN FACT, TRUE.

SO WITH THIS, THE TIME WAS APPROACHING TO WRITE A SYNTHESIS OF KNOWLEDGE ABOUT THE SOCIAL INSECTS.
I DREAMED OF SPINNING CRYSTAL-CLEAR SUMMARIES OF THEIR CLASSIFICATION, ANATOMY, LIFE CYCLES, BEHAVIOR, AND SOCIAL ORGANIZATION IN A SINGLE, WELL-ILLUSTRATED VOLUME.
ANOTHER LONG STORY SHORT—THIS ONE ABOUT SIX YEARS LONG—MY BOOK *THE INSECT SOCIETIES*, PUBLISHED IN 1971, CONVEYED MY VISION OF THE SOCIAL INSECTS.
AND IN ITS FINAL PARAGRAPH, I LOOKED TO THE FUTURE PROSPECTS OF SOCIOBIOLOGY...
"IN SPITE OF THE PHYLOGENETIC REMOTENESS OF VERTEBRATES AND INSECTS AND THE BASIC DISTINCTION BETWEEN THEIR RESPECTIVE PERSONAL AND IMPERSONAL SYSTEMS OF COMMUNICATION, THESE TWO GROUPS OF ANIMALS HAVE EVOLVED SOCIAL BEHAVIORS THAT ARE SIMILAR IN DEGREE OF COMPLEXITY AND CONVERGENT IN MANY IMPORTANT DETAILS."
"THIS FACT CONVEYS A SPECIAL PROMISE THAT SOCIOBIOLOGY CAN EVENTUALLY BE DERIVED FROM THE FIRST PRINCIPLES OF POPULATION AND BEHAVIORAL BIOLOGY AND DEVELOPED INTO A SINGLE, MATURE SCIENCE. THE DISCIPLINE CAN THEN BE EXPECTED TO INCREASE OUR UNDERSTANDING OF THE UNIQUE QUALITIES OF SOCIAL BEHAVIOR IN ANIMALS AS OPPOSED TO THOSE OF MAN."
OKAY, THAT'S DONE. WHERE MIGHT I GO NEXT?
VERTEBRATE BEHAVIOR—BIRDS, MAMMALS, REPTILES, AMPHIBIANS, FISHES—SEEMED TOO FORMIDABLE A SUBJECT TO ENTER FROM THE DIRECTION OF ENTOMOLOGY. BUT I FOUND OUT I WAS WRONG.
AFTER PROBING A BIT, TALKING WITH SPECIALISTS, I HAD A REVELATION. VERTEBRATES WEREN'T DIFFICULT AT ALL.
ENTOMOLOGY IS A TECHNICALLY MORE DIFFICULT SUBJECT, PARTLY BECAUSE INSECTS ARE SO MUCH MORE DIVERSE—OVER A MILLION KNOWN SPECIES VERSUS ABOUT 60,000 VERTEBRATES...

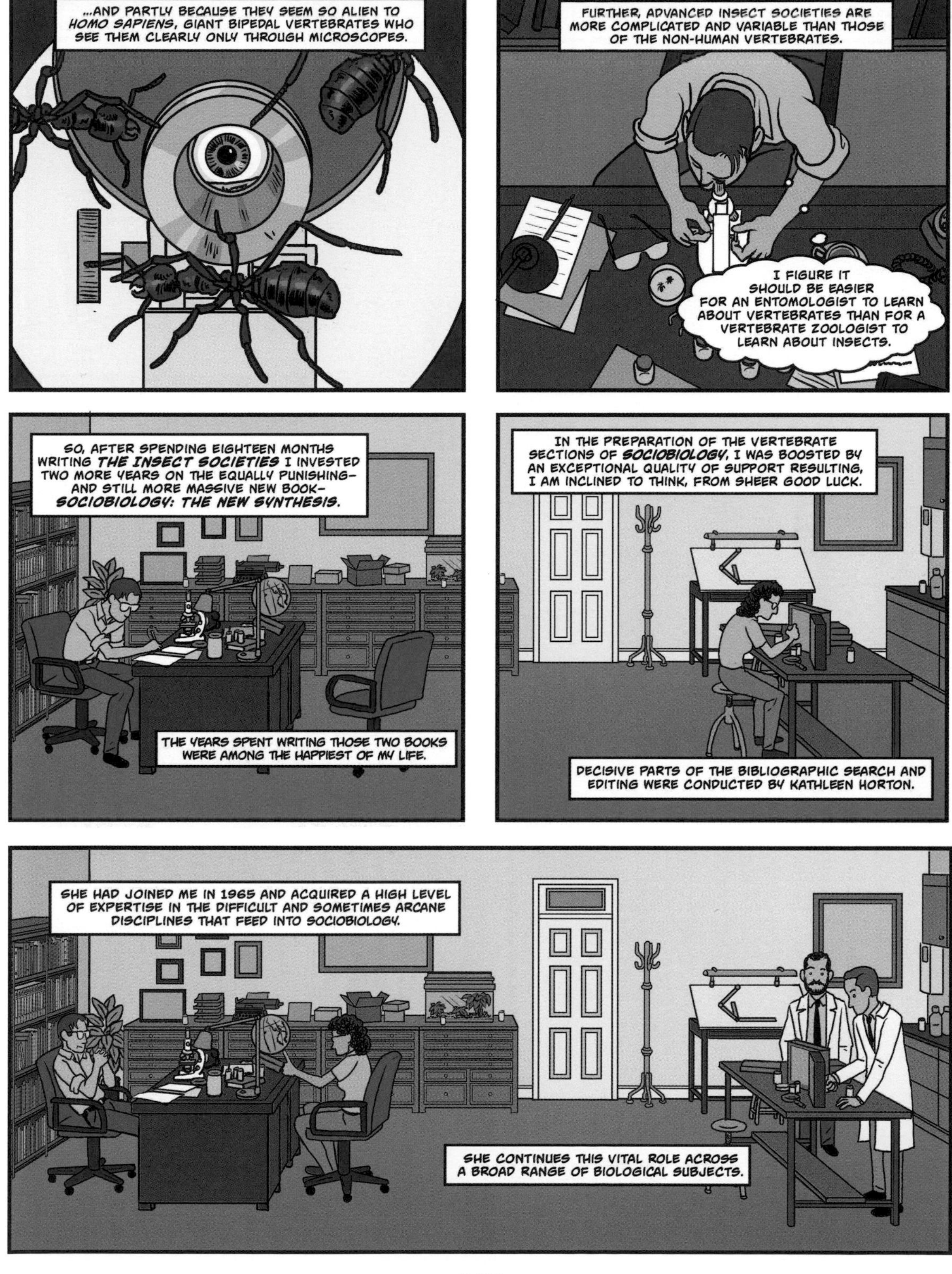
...AND PARTLY BECAUSE THEY SEEM SO ALIEN TO HOMO SAPIENS, GIANT BIPEDAL VERTEBRATES WHO SEE THEM CLEARLY ONLY THROUGH MICROSCOPES.
FURTHER, ADVANCED INSECT SOCIETIES ARE MORE COMPLICATED AND VARIABLE THAN THOSE OF THE NON-HUMAN VERTEBRATES.
I FIGURE IT SHOULD BE EASIER FOR AN ENTOMOLOGIST TO LEARN ABOUT VERTEBRATES THAN FOR A VERTEBRATE ZOOLOGIST TO LEARN ABOUT INSECTS.
SO, AFTER SPENDING EIGHTEEN MONTHS WRITING THE INSECT SOCIETIES I INVESTED TWO MORE YEARS ON THE EQUALLY PUNISHING– AND STILL MORE MASSIVE NEW BOOK– SOCIOBIOLOGY: THE NEW SYNTHESIS.
THE YEARS SPENT WRITING THOSE TWO BOOKS WERE AMONG THE HAPPIEST OF MY LIFE.
IN THE PREPARATION OF THE VERTEBRATE SECTIONS OF SOCIOBIOLOGY, I WAS BOOSTED BY AN EXCEPTIONAL QUALITY OF SUPPORT RESULTING, I AM INCLINED TO THINK, FROM SHEER GOOD LUCK.
DECISIVE PARTS OF THE BIBLIOGRAPHIC SEARCH AND EDITING WERE CONDUCTED BY KATHLEEN HORTON.
SHE HAD JOINED ME IN 1965 AND ACQUIRED A HIGH LEVEL OF EXPERTISE IN THE DIFFICULT AND SOMETIMES ARCANE DISCIPLINES THAT FEED INTO SOCIOBIOLOGY.
SHE CONTINUES THIS VITAL ROLE ACROSS A BROAD RANGE OF BIOLOGICAL SUBJECTS.

SARAH LANDRY, THEN AS NOW ONE OF AMERICA'S BEST WILDLIFE ILLUSTRATORS, WAS MIRACULOUSLY AVAILABLE IN THE EARLY PART OF HER CAREER AS I STARTED WORK ON MY BIG BOOKS.
WITH A PASSION FOR ACCURACY, SHE WENT BEYOND THE EFFORT REQUIRED FOR AN ORDINARY BOOK ON ANIMAL BEHAVIOR.
SHE TRAVELED TO ZOOS AND AQUARIA TO SKETCH CAPTIVE ANIMALS...
...AND VISITED HERBARIA TO RENDER IN DETAIL THE PLANT SPECIES FOUND IN THE NATURAL HABITATS OF THE ANIMAL SOCIETIES. TO SARAH, THE BUSHES AMONG WHICH A MOUNTAIN GORILLA FORAGED MEANT AS MUCH AS THE GORILLA ITSELF.
I ALSO BENEFITED FROM THE WORK OF ANOTHER STUDENT, ROBERT TRIVERS.
HE WAS OBLIVIOUS TO, OR UNCARING OF, THE CUSTOM OF MAKING APPOINTMENTS.

PLODIPLOID ANTS,
EQUAL NUMBERS OF SISTERS
THEY END UP WITH AN AVERAGE
IP FOR ALL THE OFFSPRING OF ONE-
ELING THE APPARENT ADVANTAGE
WE CAN EXPRESS THIS
HE CAME IN WITH A FLOOD OF IDEAS, NEW INFORMATION, AND CHALLENGES, DELIVERED IN IRONY AND MERRIMENT.
1/2 (FRACTION OF FEMALES) X 3/4 (GENES SHARED) + 1/2 (FRACTION OF MALES) X 1/4 (GENES SHARED) = 1/2 BUT...
WAIT A MINU...
ONE-HALF OF THE GENES SHARED, ON AVERAGE, IS THE SAME PAYOFF AS FROM THE ORDINARY PRODUCTION OF SONS AND DAUGHTERS WITHOUT HAPLODIPLOIDY. ONLY IF THE WORKERS CAN RAISE A HIGHER PERCENTAGE OF SISTERS IN THE ROYAL BROOD CAN THEY REAP THE LARGER REWARDS OF ALTRUISM TWISTED BY HAPLODIPLOIDY.
THE BEST POSSIBLE OVERALL DEGREE OF RELATIONSHIP IN THE HYMENOPTERA IS FIVE-EIGHTHS, REACHED BY INVESTING THREE-FOURTHS OF THE RESOURCES IN SISTERS. SO... 3/4 (FRACTION OF FEMALES) X 3/4 (GENES SHARED) + 1/4 (FRACTION OF MALES) X 1/4 (GENES SHARED) = 5/8
5/8 BEATS 1/2 AND GIVES THE ADVANTAGE TO COLONIAL EXISTENCE, IF ALL OTHER CONDITIONS ARE EQUAL.
MY OWN PLEASURE IN THESE EXCHANGES WAS TINGED WITH A SENSE OF PSYCHOLOGICAL RISK, AS THOUGH TESTING A MIND-ALTERING AND POSSIBLY DANGEROUS DRUG.
AS IF CONDITIONS WERE EVER EQUAL. HAVE I TOLD YOU ABOUT OUR FUNDING SITUATION?
HE WAS OFTEN DAZZLING, AND OUR SCIENCE WAS ADVANCED BY HILARITY AS WE SWITCHED FROM CONCEPT TO GOSSIP TO JOKE AND BACK TO CONCEPT.

1/2 (FRACTION OF FEMALES) X 3/4 (GENES SHARED) + 1/2 (FRACTION OF MALES) X 1/4 (GENES SHARED) = 1/2
TWO OR THREE HOURS WITH TRIVERS LEFT ME EXHAUSTED FOR THE DAY, BUT FOR A FEW SPECTACULAR YEARS, 1971 THROUGH 1974, TRIVERS BLAZED NEW PATHS IN SOCIOBIOLOGICAL THEORY.
HE GENERATED A MODEL OF RECIPROCAL ALTRUISM, BY WHICH HUMANS AND MORE INTELLIGENT ANIMALS EVOLVE CONTRACT RULES THAT REACH BEYOND SELF-SACRIFICE BASED ON KIN SELECTION.
HIS MOST IMPORTANT CONTRIBUTION, THE THEORY OF THE FAMILY, SET THE FOUNDATION FOR TODAY'S RESEARCH ON THESE SUBJECTS WITHIN BEHAVIORAL BIOLOGY.
IT'S COMPLICATED, OF COURSE—FAMILIES ALWAYS ARE—BUT TRIVERS PROVIDED A PLAUSIBLE ARGUMENT FOR THE ULTIMATE CAUSATION OF CONFLICT.
AND SO WE ARRIVED AT *SOCIOBIOLOGY: THE NEW SYNTHESIS.*
Sociobiology
THE NEW SYNTHESIS
Edward O. Wilson
IN IT, I COVERED ALL ORGANISMS THAT COULD EVEN REMOTELY BE CALLED SOCIAL, FROM COLONIAL BACTERIA AND AMOEBAE TO TROOPS OF MONKEYS AND OTHER PRIMATES.

CORALS, SIPHONOPHORES, AND OTHER INVERTEBRATES
I RECOGNIZED FOUR "PINNACLES" OF SOCIAL EVOLUTION.
THEIR SOCIETIES ARE, FIRST, INDEPENDENTLY DERIVED IN EVOLUTION.
SOCIAL INSECTS
SECOND, THEY ARE COMPLEX OR SOPHISTICATED IN ORGANIZATION.
SOCIAL VERTEBRATES, ESPECIALLY THE GREAT APES AND OTHER OLD-WORLD PRIMATES
AND, FINALLY, POSSESSED OF GENETIC STRUCTURES AND ORGANIZATIONS DIFFERING RADICALLY FROM THOSE OF THE OTHERS.
THE FOURTH IS MAN.
YES, MAN.
Sociobiology
THE NEW SYNTHESIS
Edward O. Wilson
I DID NOT HESITATE TO INCLUDE *HOMO SAPIENS*, BECAUSE NOT TO HAVE DONE SO WOULD HAVE BEEN TO OMIT A MAJOR PART OF BIOLOGY.

WE ARE, AFTER ALL, A BIOLOGICAL SPECIES. OUR HISTORY DID NOT BEGIN 10,000 YEARS AGO IN THE VILLAGES OF ANATOLIA AND JORDAN.
IT SPANS THE 2 MILLION YEARS OF THE LIFE OF THE GENUS *HOMO*. DEEP HISTORY–BIOLOGICAL HISTORY–MADE US WHAT WE ARE, NO LESS THAN CULTURE.
SO, IT FELT APPROPRIATE TO USE PROVOCATIVE LANGUAGE AS I OPENED THE FINAL CHAPTER OF *SOCIOBIOLOGY*. HERE IS WHAT I SAID...
"LET US NOW CONSIDER MAN IN THE FREE SPIRIT OF NATURAL HISTORY, AS THOUGH WE WERE ZOOLOGISTS FROM ANOTHER PLANET COMPLETING A CATALOG OF SOCIAL SPECIES ON EARTH.
"IN THIS MACROSCOPIC VIEW, THE HUMANITIES AND SOCIAL SCIENCES SHRINK TO SPECIALIZED BRANCHES OF BIOLOGY.
"HISTORY, BIOGRAPHY, AND FICTION ARE THE RESEARCH PROTOCOLS OF HUMAN ETHOLOGY.
"AND ANTHROPOLOGY AND SOCIOLOGY TOGETHER CONSTITUTE THE SOCIOBIOLOGY OF A SINGLE PRIMATE SPECIES."

PERHAPS I SHOULD HAVE STOPPED AT CHIMPANZEES.
MANY BIOLOGISTS WISH I HAD.
REGARDLESS OF THE BOOK'S REAL STRENGTH, MUCH OF THE CONTROVERSY MIGHT HAVE BEEN AVOIDED, AND FOR THAT I MUST BEAR THE RESPONSIBILITY.
I WROTE *SOCIOBIOLOGY: THE NEW SYNTHESIS* AS TWO DIFFERENT BOOKS IN ONE.
THE FIRST TWENTY-SIX CHAPTERS—94% OF THE TEXT—WAS AN ENCYCLOPEDIC REVIEW OF SOCIAL MICROORGANISMS AND ANIMALS, ORGANIZED ACCORDING TO THE PRINCIPLES OF EVOLUTIONARY THEORY.
THE SECOND BOOK—THE LAST CHAPTER OF TWENTY-NINE DOUBLE-COLUMNED PAGES—CONSISTED MOSTLY OF FACTS FROM THE SOCIAL SCIENCES, INTERPRETED BY HYPOTHESES ON THE BIOLOGICAL FOUNDATIONS OF HUMAN BEHAVIOR.
THE DIFFERENCES BETWEEN BOOKS ONE AND TWO GIVE RISE TO THE DUAL SOCIOBIOLOGIES OF POPULAR PERCEPTION.
THUNK!
THE FIRST IS SOCIOBIOLOGY AS I INTENDED TO PORTRAY IT—THE SYSTEMATIC STUDY OF THE BIOLOGICAL BASIS OF SOCIAL BEHAVIOR AND ADVANCED SOCIETIES...
AND THEN THERE IS THE EVIL TWIN AS PERCEIVED BY SO MANY—THE SCIENTIFIC-IDEOLOGICAL DOCTRINE THAT HUMAN SOCIAL BEHAVIOR IS DETERMINED BY GENES.

PEOPLE HATE GENETIC DETERMINISM.
BUT, THERE'S A GOOD ARGUMENT FOR IT. HERE IS MINE...

HUMANS INHERIT A PROPENSITY TO ACQUIRE BEHAVIOR AND SOCIAL STRUCTURES, A PROPENSITY SHARED BY ENOUGH PEOPLE THAT WE CALL IT "HUMAN NATURE."
NOW, INDIVIDUALS HAVE FREE WILL AND THE CHOICE TO TURN IN MANY DIRECTIONS.

BUT THE CHANNELS OF OUR PSYCHOLOGICAL DEVELOPMENT ARE NEVERTHELESS—HOWEVER MUCH WE MIGHT WISH OTHERWISE—CUT MORE DEEPLY BY THE GENES IN CERTAIN DIRECTIONS.
SO, WHILE CULTURES VARY GREATLY, THEY CONVERGE TOWARD THESE TRAITS.
THE MANHATTANITE AND NEW GUINEA HIGHLANDER HAVE BEEN SEPARATED BY 50,000 YEARS OF HISTORY.

BUT THEY...WE...STILL UNDERSTAND EACH OTHER, FOR THE ELEMENTARY REASON THAT OUR COMMON HUMANITY IS PRESERVED IN THE GENES WE SHARE FROM OUR COMMON ANCESTRY.

IN *SOCIOBIOLOGY*, I CONJECTURED THAT THERE MIGHT BE SINGLE, STILL UNIDENTIFIED GENES AFFECTING AGGRESSION, ALTRUISM, AND OTHER BEHAVIORS.

I WAS WELL AWARE THAT SUCH TRAITS ARE USUALLY CONTROLLED BY MULTIPLE GENES, AND THAT ENVIRONMENT PLAYS A MAJOR ROLE IN CREATING VARIATION AMONG INDIVIDUALS AND SOCIETIES.
YET, WHATEVER THE EXACT NATURE OF THE GENETIC CONTROLS, I CONTENDED, THE IMPORTANT POINT IS THAT HEREDITY INTERACTS WITH ENVIRONMENT TO CREATE A GRAVITATIONAL PULL TOWARD A FIXED MEAN.
IT GATHERS PEOPLE IN ALL SOCIETIES INTO THAT NARROW STATISTICAL CIRCLE THAT WE DEFINE AS HUMAN NATURE.
MINE WAS AN EXCEPTIONALLY STRONG HEREDITARIAN POSITION FOR THE TIME.
THE VIEW FAVORED BY MOST SOCIAL THEORISTS, THAT HUMAN NATURE IS BUILT WHOLLY FROM EXPERIENCE, WAS NOT JUST ANOTHER HYPOTHESIS UP FOR TESTING. IN THE 1970S, IT WAS A DEEPLY ROOTED PHILOSOPHY.
THE HYPOTHESIS THAT HUMAN NATURE HAS A *GENETIC* FOUNDATION CALLED MANY ASSUMPTIONS INTO QUESTION.
Sociobiology
IF, IN FACT, GENES DID SURRENDER THEIR CONTROL SOMETIME BACK DURING HUMAN EVOLUTION, BIOLOGY CAN PLAY NO CONTRIBUTORY ROLE IN THE SOCIAL SCIENCES.
AND IF HUMAN NATURE IS MOSTLY ACQUIRED, AND NO SIGNIFICANT PART OF IT IS INHERITED, THEN IT IS ***EASIER*** TO CONCLUDE THAT DIFFERENT CULTURES MUST BE ACCORDED MORAL EQUIVALENCY, AND DESERVE RESPECT.
FOR WHAT IS THOUGHT GOOD AND TRUE HAS BEEN DETERMINED MORE BY POWER THAN BY INTRINSIC VALIDITY.
THE CULTURES OF OPPRESSED PEOPLES ARE TO BE SPECIALLY VALUED, BECAUSE THE HISTORIES OF CULTURAL CONFLICT ARE WRITTEN BY THE VICTORS.
THE HYPOTHESIS THAT HUMAN NATURE HAS A GENETIC FOUNDATION CALLED THESE ASSUMPTIONS INTO QUESTION.

TRIBALISM AND GENDER DIFFERENCES MIGHT THEN BE JUDGED UNAVOIDABLE, AND CLASS DIFFERENCES AND WAR IN SOME MANNER MIGHT SEEM "NATURAL."
AND THAT WOULD BE JUST THE BEGINNING.
BECAUSE PEOPLE UNQUESTIONABLY VARY IN HEREDITARY PHYSICAL TRAITS, THEY MIGHT ALSO DIFFER IRREVERSIBLY IN PERSONAL ABILITY AND EMOTIONAL ATTRIBUTES.
IN THE 1970S, A GREAT MANY ORDINARY PEOPLE BELIEVED THESE HEREDITARIAN PROPOSITIONS TO BE MORE OR LESS TRUE. BUT...
...ANYONE WHO ADVANCED SUCH IDEAS IN COLLEGES AND UNIVERSITIES RISKED THE SCALDING CHARGES OF RACISM AND SEXISM.
IN CONTRAST, THOSE WHO ATTACKED THE HEREDITARIAN POSITION WERE PRAISED AS DEFENDERS OF TRUTH AND VIRTUE.
AND I WAS IN A UNIVERSITY.

TO MY MIND, I HAD SIMPLY EXPANDED THE RANGE OF SUBJECTS THAT INTERESTED ME, STARTING WITH ANTS AND PROCEEDING TO SOCIAL INSECTS, THEN TO ANIMALS, AND FINALLY TO HUMANITY.
BELIEVING THE TIME RIPE FOR MELDING BIOLOGY AND THE SOCIAL SCIENCES, I CAME TO BELIEVE THAT EVOLUTIONARY BIOLOGY SHOULD SERVE AS THE FOUNDATION OF THE SOCIAL SCIENCES.
RIP!
WHAT MADE *SOCIOBIOLOGY* NOTORIOUS, THEN, WAS ITS HYBRID NATURE.
HAD THE TWO PARTS OF THE BOOK BEEN PUBLISHED SEPARATELY, THE BIOLOGICAL CORE WOULD HAVE BEEN WELL RECEIVED BY SPECIALISTS IN ANIMAL BEHAVIOR AND ECOLOGY, WHILE THE WRITINGS ON HUMAN BEHAVIOR MIGHT EASILY HAVE BEEN DISMISSED OR IGNORED.
PLACED BETWEEN THE SAME TWO COVERS, HOWEVER, THE WHOLE WAS GREATER THAN THE SUM OF ITS PARTS.
THE HUMAN CHAPTERS WERE RENDERED CREDITABLE BY THE MASSIVE ANIMAL DOCUMENTATION, WHILE THE BIOLOGY GAINED ADDED SIGNIFICANCE FROM THE HUMAN IMPLICATIONS.
OR SO I THOUGHT. BUT THE CONJUNCTION CREATED A SYLLOGISM THAT PROVED UNPALATABLE TO MANY.
SOCIOBIOLOGY IS PART OF BIOLOGY. ^ BIOLOGY IS RELIABLE. => HUMAN SOCIOBIOLOGY IS RELIABLE.
I USED STRONG, PROVOCATIVE LANGUAGE TO START THE PROCESS.

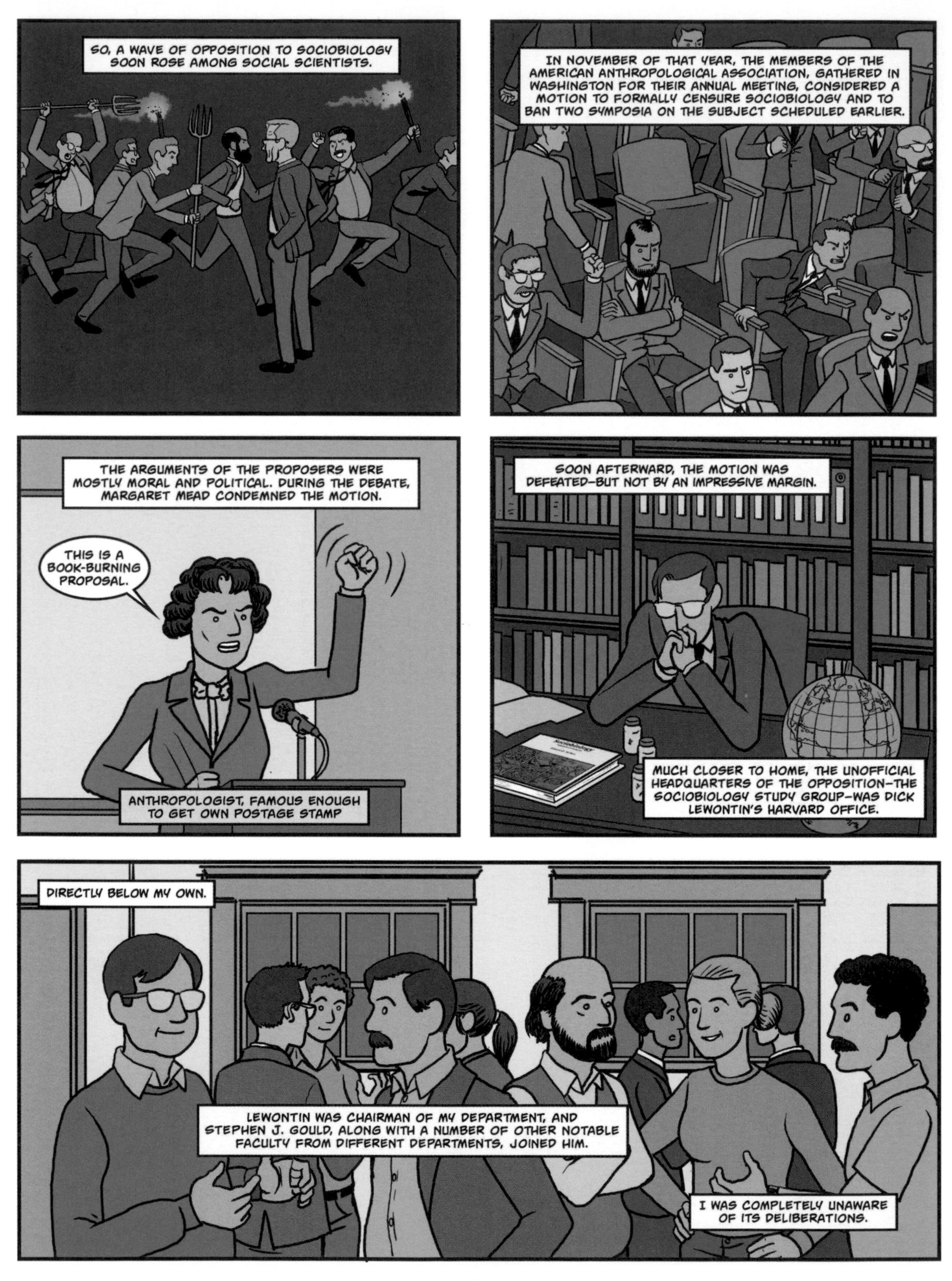
SO, A WAVE OF OPPOSITION TO SOCIOBIOLOGY SOON ROSE AMONG SOCIAL SCIENTISTS.
IN NOVEMBER OF THAT YEAR, THE MEMBERS OF THE AMERICAN ANTHROPOLOGICAL ASSOCIATION, GATHERED IN WASHINGTON FOR THEIR ANNUAL MEETING, CONSIDERED A MOTION TO FORMALLY CENSURE SOCIOBIOLOGY AND TO BAN TWO SYMPOSIA ON THE SUBJECT SCHEDULED EARLIER.
THE ARGUMENTS OF THE PROPOSERS WERE MOSTLY MORAL AND POLITICAL. DURING THE DEBATE, MARGARET MEAD CONDEMNED THE MOTION.
THIS IS A BOOK-BURNING PROPOSAL.
ANTHROPOLOGIST, FAMOUS ENOUGH TO GET OWN POSTAGE STAMP
SOON AFTERWARD, THE MOTION WAS DEFEATED—BUT NOT BY AN IMPRESSIVE MARGIN.
MUCH CLOSER TO HOME, THE UNOFFICIAL HEADQUARTERS OF THE OPPOSITION—THE SOCIOBIOLOGY STUDY GROUP—WAS DICK LEWONTIN'S HARVARD OFFICE.
DIRECTLY BELOW MY OWN.
LEWONTIN WAS CHAIRMAN OF MY DEPARTMENT, AND STEPHEN J. GOULD, ALONG WITH A NUMBER OF OTHER NOTABLE FACULTY FROM DIFFERENT DEPARTMENTS, JOINED HIM.
I WAS COMPLETELY UNAWARE OF ITS DELIBERATIONS.

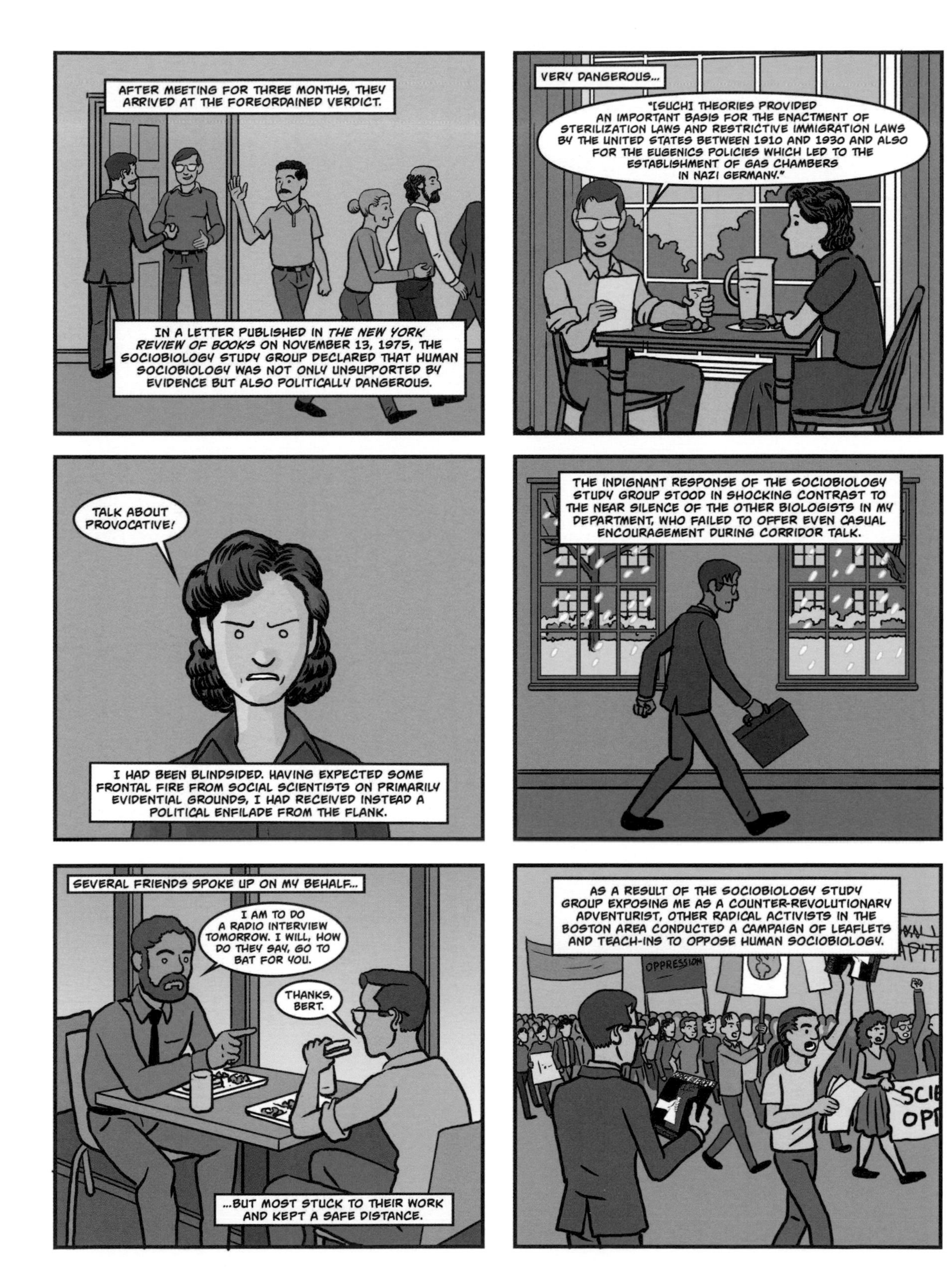
AFTER MEETING FOR THREE MONTHS, THEY ARRIVED AT THE FOREORDAINED VERDICT.
IN A LETTER PUBLISHED IN *THE NEW YORK REVIEW OF BOOKS* ON NOVEMBER 13, 1975, THE SOCIOBIOLOGY STUDY GROUP DECLARED THAT HUMAN SOCIOBIOLOGY WAS NOT ONLY UNSUPPORTED BY EVIDENCE BUT ALSO POLITICALLY DANGEROUS.
VERY DANGEROUS...
"[SUCH] THEORIES PROVIDED AN IMPORTANT BASIS FOR THE ENACTMENT OF STERILIZATION LAWS AND RESTRICTIVE IMMIGRATION LAWS BY THE UNITED STATES BETWEEN 1910 AND 1930 AND ALSO FOR THE EUGENICS POLICIES WHICH LED TO THE ESTABLISHMENT OF GAS CHAMBERS IN NAZI GERMANY."
TALK ABOUT PROVOCATIVE!
I HAD BEEN BLINDSIDED. HAVING EXPECTED SOME FRONTAL FIRE FROM SOCIAL SCIENTISTS ON PRIMARILY EVIDENTIAL GROUNDS, I HAD RECEIVED INSTEAD A POLITICAL ENFILADE FROM THE FLANK.
THE INDIGNANT RESPONSE OF THE SOCIOBIOLOGY STUDY GROUP STOOD IN SHOCKING CONTRAST TO THE NEAR SILENCE OF THE OTHER BIOLOGISTS IN MY DEPARTMENT, WHO FAILED TO OFFER EVEN CASUAL ENCOURAGEMENT DURING CORRIDOR TALK.
SEVERAL FRIENDS SPOKE UP ON MY BEHALF...
I AM TO DO A RADIO INTERVIEW TOMORROW. I WILL, HOW DO THEY SAY, GO TO BAT FOR YOU.
THANKS, BERT.
...BUT MOST STUCK TO THEIR WORK AND KEPT A SAFE DISTANCE.
AS A RESULT OF THE SOCIOBIOLOGY STUDY GROUP EXPOSING ME AS A COUNTER-REVOLUTIONARY ADVENTURIST, OTHER RADICAL ACTIVISTS IN THE BOSTON AREA CONDUCTED A CAMPAIGN OF LEAFLETS AND TEACH-INS TO OPPOSE HUMAN SOCIOBIOLOGY.
OPPRESSION

I BRIEFLY CONSIDERED OFFERS OF PROFESSORSHIPS FROM THREE UNIVERSITIES—IN CASE, THEIR REPRESENTATIVES SAID, I WISHED TO LEAVE THE PHYSICAL CENTER OF THE CONTROVERSY. BUT IT ALL CAME TO VERY LITTLE.
SCIENCE FOR THE PEOPLE

AND BECAUSE OF MY RESPECT FOR THE MEMBERS OF THE SOCIOBIOLOGY STUDY GROUP I KNEW PERSONALLY, I WAS AT FIRST STRUCK BY SELF-DOUBT.
HAVE I TAKEN A FATAL INTELLECTUAL MISSTEP BY CROSSING THE LINE INTO HUMAN BEHAVIOR?

THEN I RETHOUGHT MY OWN EVIDENCE AND LOGIC.
...
HELL NO. WHAT I SAID IS DEFENSIBLE AS SCIENCE. THE ATTACK IS POLITICAL, NOT EVIDENTIAL.

AS MY MIND SETTLED ON THE DETAILS, ANGER REPLACED ANXIETY. I PENNED AN INDIGNANT REBUTTAL FOR *THE NEW YORK REVIEW OF BOOKS*.
IN A FEW MORE WEEKS, MY OLD CONFIDENCE RETURNED, REPLACING THE ANGER WITH A FRESH SURGE OF AMBITION.

THERE WAS AN ENEMY IN THE FIELD. AN IMPORTANT ENEMY. AND A NEW SUBJECT—WHICH, FOR ME, MEANT OPPORTUNITY...

...WHICH CONTRIBUTED TO MY BEING AWARDED THE NATIONAL MEDAL OF SCIENCE IN 1976 AND LED TO *ON HUMAN NATURE*, WHICH WON THE 1979 PULITZER PRIZE FOR GENERAL NONFICTION.

LITERARY AND POLITICAL, NOT SCIENTIFIC, VALIDATION.
HEY, AFTER ALL THIS NONSENSE, ANY PORT IN A STORM.
THE STORM WASN'T QUITE OVER, AND THE MOST DRAMATIC EPISODE IN THE STORY WAS STILL A FRESH MEMORY.
ON FEBRUARY 15, 1978, I HAD GONE TO SPEAK AT A SYMPOSIUM ON SOCIOBIOLOGY PLANNED AS PART OF THE ANNUAL MEETING OF THE LARGEST ORGANIZATION OF SCIENTISTS IN THE WORLD.
AAAS
ADVANCING SCIENCE, SERVING SOCIETY
CONFERENCE
THE AMERICAN ASSOCIATION FOR THE ADVANCEMENT OF SCIENCE WAS AND REMAINS ESPECIALLY CONCERNED WITH THE RELATION OF SCIENCE TO EDUCATION AND PUBLIC POLICY.
A LARGE CROWD WAS EXPECTED AT THE SYMPOSIUM, WHICH FEATURED A HALF-DOZEN OF THE PRINCIPAL RESEARCHERS ON HUMAN SOCIOBIOLOGY, AS WELL AS ONE OF ITS MOST ARTICULATE CRITICS, STEPHEN JAY GOULD.

WAR
RACISM
SCIENCE = PROFITS
AAA$
NEXT IS PROFESSOR E.O. WILSON, WITH HIS TALK TITLED "WHAT IS SOCIOBIOLOGY."

I THINK WE ALL KNOW HIM AND HIS WORK, BUT WHAT YOU DON'T KNOW IS HE FRACTURED HIS ANKLE WHILE JOGGING RECENTLY.

SO WE'RE NOT GOING TO ASK HIM TO COME TO THE LECTERN. INSTEAD...
RACISM

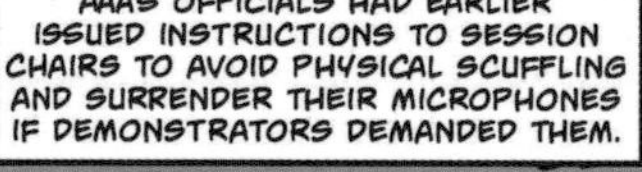
AAAS OFFICIALS HAD EARLIER ISSUED INSTRUCTIONS TO SESSION CHAIRS TO AVOID PHYSICAL SCUFFLING AND SURRENDER THEIR MICROPHONES IF DEMONSTRATORS DEMANDED THEM.

WE, THE INTERNATIONAL COMMITTEE AGAINST RACISM, DEMAND A VOICE AGAINST THIS FASC...
FINE. BUT IF THE MICROPHONES ARE NOT RETURNED WITHIN TWO MINUTES, HOTEL SECURITY WILL BE CALLED.

DR. E.O. WILSON

WILSON, YOU'RE ALL WET!
WILSON, YOU'RE ALL WET!
WILSON, YOU'RE ALL WET!
DR. E.O. WILSON

DR. E.O. WILSON

THIS IS WHAT IT TAKES TO GET APPLAUSE?
OF COURSE, WHAT ELSE CAN YOU ALL DO? YOU MIGHT BE NEXT.

BEFORE I COULD PROCEED WITH MY BRIEF LECTURE, OTHER MEMBERS OF THE PANEL ROSE TO CONDEMN THE INCAR ACTION.
VIOLENCE AS RADICAL POSTURING, TO QUOTE LENIN, IS AN "INFANTILE DISORDER" OF SOCIALISM.
YOU'RE RIGHT. BUT IT'S THE GROWN-UP INTELLECTUALS THAT I HAVE TO WORRY ABOUT.

HOW DID I FEEL DURING THE INCIDENT?
CALM—DARE I SAY ICY COLD?—AS I LET THE PROTESTORS' ANGER WASH OVER ME.

SO, INFAMOUS ...IN A SMALL WAY.
IT SHOULD HAVE BEEN OBVIOUS TO ME THAT HUMAN SOCIOBIOLOGY WOULD REMAIN IN TROUBLE, BOTH INTELLECTUALLY AND POLITICALLY, UNTIL IT INCORPORATED CULTURE INTO ITS ANALYSES.
AND NO ONE CAN BE SURE OF ANYTHING UNTIL AN ATTEMPT IS MADE.
I REMAIN CONVINCED THAT THE TRUE NATURE OF GENE-CULTURE COEVOLUTION IS THE CENTRAL PROBLEM OF THE SOCIAL SCIENCES, AND ONE OF THE GREAT UNEXPLORED DOMAINS OF SCIENCE GENERALLY.
I THINK IT'S TIME WILL COME...

PART THREE
Biodiversity, Biophilia

A MAN'S WORK IS NOTHING BUT THIS SLOW TREK
TO REDISCOVER, THROUGH THE DETOURS OF ART,
THOSE TWO OR THREE GREAT AND SIMPLE IMAGES
IN WHOSE PRESENCE HIS HEART FIRST OPENED.

[ALBERT CAMUS]

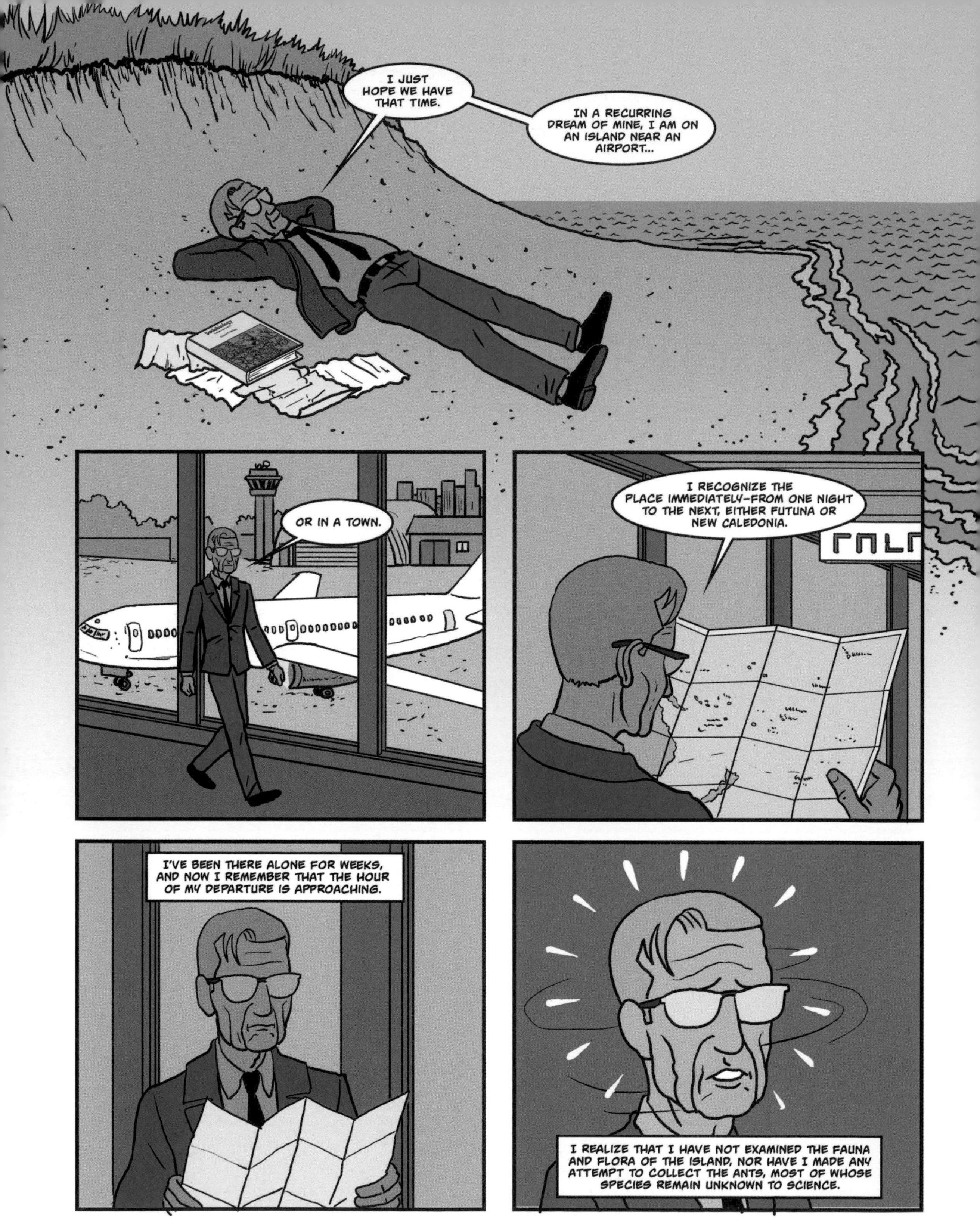
I JUST HOPE WE HAVE THAT TIME.
IN A RECURRING DREAM OF MINE, I AM ON AN ISLAND NEAR AN AIRPORT...
OR IN A TOWN.
I RECOGNIZE THE PLACE IMMEDIATELY—FROM ONE NIGHT TO THE NEXT, EITHER FUTUNA OR NEW CALEDONIA.
I'VE BEEN THERE ALONE FOR WEEKS, AND NOW I REMEMBER THAT THE HOUR OF MY DEPARTURE IS APPROACHING.
I REALIZE THAT I HAVE NOT EXAMINED THE FAUNA AND FLORA OF THE ISLAND, NOR HAVE I MADE ANY ATTEMPT TO COLLECT THE ANTS, MOST OF WHOSE SPECIES REMAIN UNKNOWN TO SCIENCE.

I BEGIN A FRANTIC SEARCH FOR NATIVE FOREST.

IN THE DISTANCE, I SEE WHAT LOOKS LIKE THE EDGE OF A COPSE AND RUN TO IT, ONLY TO FIND A ROW OF INVASIVE TREES PLANTED AS A WINDBREAK.

NOW I AM IN AN AUTOMOBILE. THERE ARE MOUNTAINS FAR TO THE NORTH.
IN EVERY DREAM, ALWAYS TO THE NORTH.

PERHAPS SOME PRISTINE FOREST REMAINS IN THE MOUNTAINS.

BUT I CANNOT GO—MY TIME HAS RUN OUT

THE DREAM ENDS, AND I AWAKEN KNOTTED WITH ANXIETY AND REGRET.

IN 1980, THE EDITORS OF *HARVARD MAGAZINE* ASKED SEVEN PROFESSORS TO IDENTIFY WHAT THEY CONSIDERED TO BE THE MOST IMPORTANT PROBLEM FACING THE WORLD IN THE COMING DECADE.
CHRONIC HUNGER AND MALNUTRITION.
THE ENORMOUSLY UNEVEN DISTRIBUTION OF WEALTH.
DOMESTIC AND SOCIAL PROBLEMS CAUSED, EXACERBATED, OR MAINTAINED BY GOVERNMENTAL BUREAUCRACIES AND POLITICAL LEADERS.
WAGE-PRICE CONTROLS.
CONTROL OF NUCLEAR WEAPONS.
MASS POVERTY.
ROBERT COLES, PROFESSOR OF PSYCHIATRY AND MEDICAL HUMANITIES
GEORGE RUPP, DEAN OF THE FACULTY OF DIVINITY
ROBERT NOZICK, PROFESSOR OF PHILOSOPHY
STEPHEN A. MARGLIN, PROFESSOR OF ECONOMICS
DAVID RIESMAN, PROFESSOR OF SOCIAL SCIENCES
PETER ROGERS, PROFESSOR OF ENVIRONMENTAL ENGINEERING, AND CITY AND REGIONAL PLANNING

AS THE ONLY NATURAL SCIENTIST, I CHOSE A RADICALLY DIFFERENT SUBJECT, AND A BROADER TIME SCALE.
MASS EXTINCTION.
THE WORST THING THAT CAN HAPPEN...
...THAT *WILL* HAPPEN, IS NOT ENERGY DEPLETION, ECONOMIC COLLAPSE, LIMITED NUCLEAR WAR, OR CONQUEST BY A TOTALITARIAN GOVERNMENT.
AS TERRIBLE AS THESE CATASTROPHES WOULD BE FOR US, THEY CAN BE REPAIRED WITHIN A FEW GENERATIONS.
THE ONE ONGOING PROCESS THAT WILL TAKE MILLIONS OF YEARS TO CORRECT IS THE LOSS OF GENETIC AND SPECIES DIVERSITY BY THE DESTRUCTION OF NATURAL HABITATS.
THIS IS THE FOLLY OUR DESCENDANTS ARE LEAST LIKELY TO FORGIVE US.
AND THIS MARKED MY DEBUT AS AN ENVIRONMENTAL ACTIVIST.

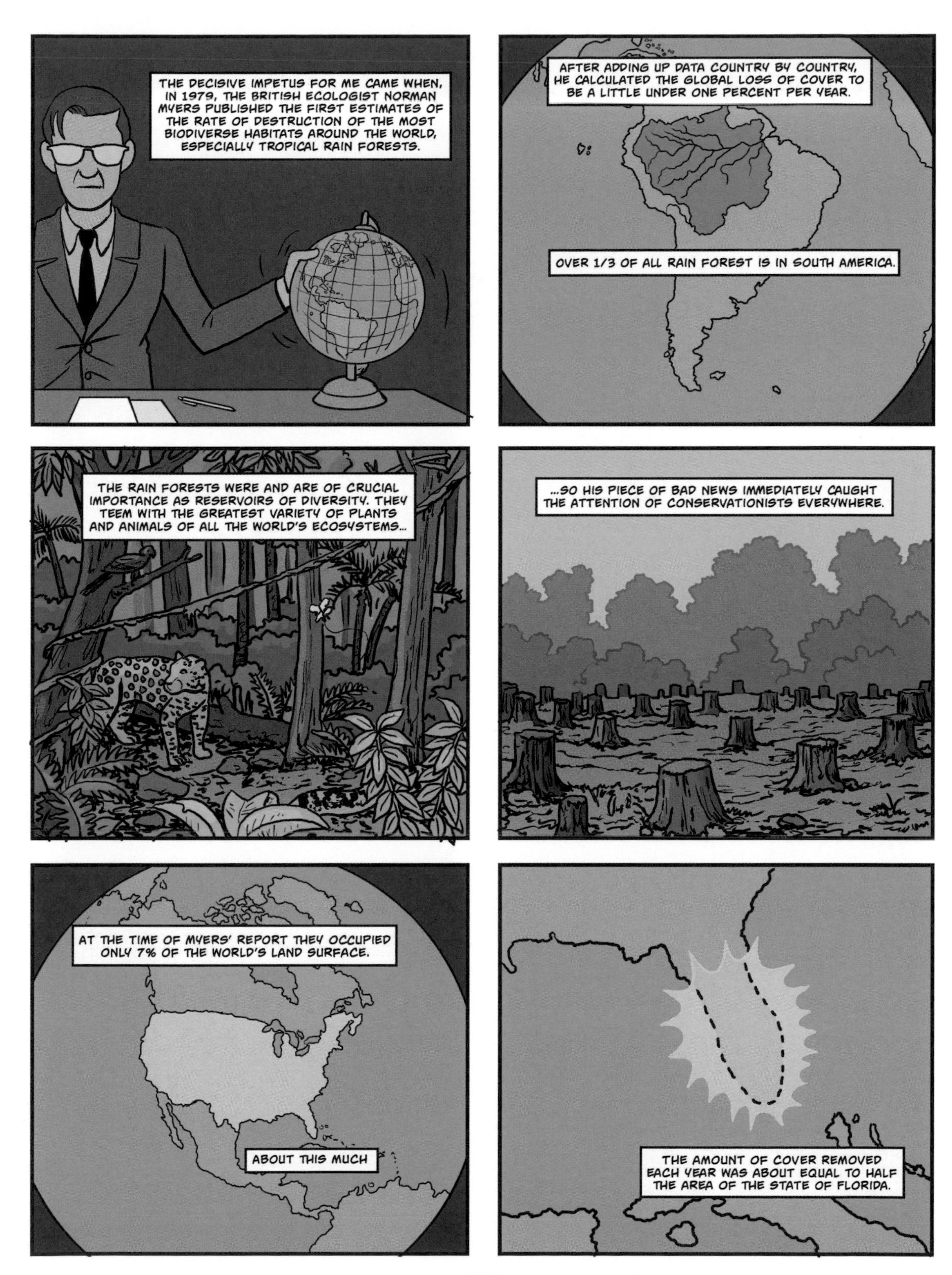
THE DECISIVE IMPETUS FOR ME CAME WHEN, IN 1979, THE BRITISH ECOLOGIST NORMAN MYERS PUBLISHED THE FIRST ESTIMATES OF THE RATE OF DESTRUCTION OF THE MOST BIODIVERSE HABITATS AROUND THE WORLD, ESPECIALLY TROPICAL RAIN FORESTS.
AFTER ADDING UP DATA COUNTRY BY COUNTRY, HE CALCULATED THE GLOBAL LOSS OF COVER TO BE A LITTLE UNDER ONE PERCENT PER YEAR.
OVER 1/3 OF ALL RAIN FOREST IS IN SOUTH AMERICA.
THE RAIN FORESTS WERE AND ARE OF CRUCIAL IMPORTANCE AS RESERVOIRS OF DIVERSITY. THEY TEEM WITH THE GREATEST VARIETY OF PLANTS AND ANIMALS OF ALL THE WORLD'S ECOSYSTEMS...
...SO HIS PIECE OF BAD NEWS IMMEDIATELY CAUGHT THE ATTENTION OF CONSERVATIONISTS EVERYWHERE.
AT THE TIME OF MYERS' REPORT THEY OCCUPIED ONLY 7% OF THE WORLD'S LAND SURFACE.
ABOUT THIS MUCH
THE AMOUNT OF COVER REMOVED EACH YEAR WAS ABOUT EQUAL TO HALF THE AREA OF THE STATE OF FLORIDA.

EACH YEAR. AND THESE DAYS WE'RE DOWN TO UNDER 3% COVER WORLDWIDE.

PRIMED BY MYERS' REPORT, I WAS FINALLY TIPPED INTO ACTIVE ENGAGEMENT BY THE EXAMPLE OF MY FRIEND PETER RAVEN.
MYER REPORT

INCREASINGLY A PUBLIC FIGURE, PETER WAS DETERMINED AND FEARLESS. HE HAD NO QUALMS ABOUT ACTIVISM...
MORE THAN ANYONE ELSE, HE MADE IT CLEAR THAT SCIENTISTS IN UNIVERSITIES AND OTHER RESEARCH-ORIENTED INSTITUTIONS MUST GET INVOLVED.
SICK OF
POLLUT
SAVE THE PLANET!
SCIENTIST AND DIRECTOR EMERITUS OF THE MISSOURI BOTANICAL GARDEN

CONSERVATION PROFESSIONALS COULD NOT BE EXPECTED TO CARRY THE BURDEN ALONE.
ONE DAY, ON IMPULSE, I CROSSED THE LINE.
PETER, I WANT YOU TO KNOW THAT I'M JOINING YOU IN THIS.

EARTH FIRST
I'M GOING TO DO EVERYTHING IN MY POWER TO HELP.

A LOOSE CONFEDERATION OF SENIOR BIOLOGISTS THAT I JOKINGLY CALLED THE "RAIN FOREST MAFIA" HAD FORMED, INCLUDING...
JARED DIAMOND, AUTHOR OF GUNS, GERMS, AND STEEL AND COLLAPSE
PAUL EHRLICH, AUTHOR OF THE POPULATION BOMB
THOMAS LOVEJOY, THE GODFATHER OF BIODIVERSITY
WE REMAINED IN FREQUENT COMMUNICATION FROM THEN ON.

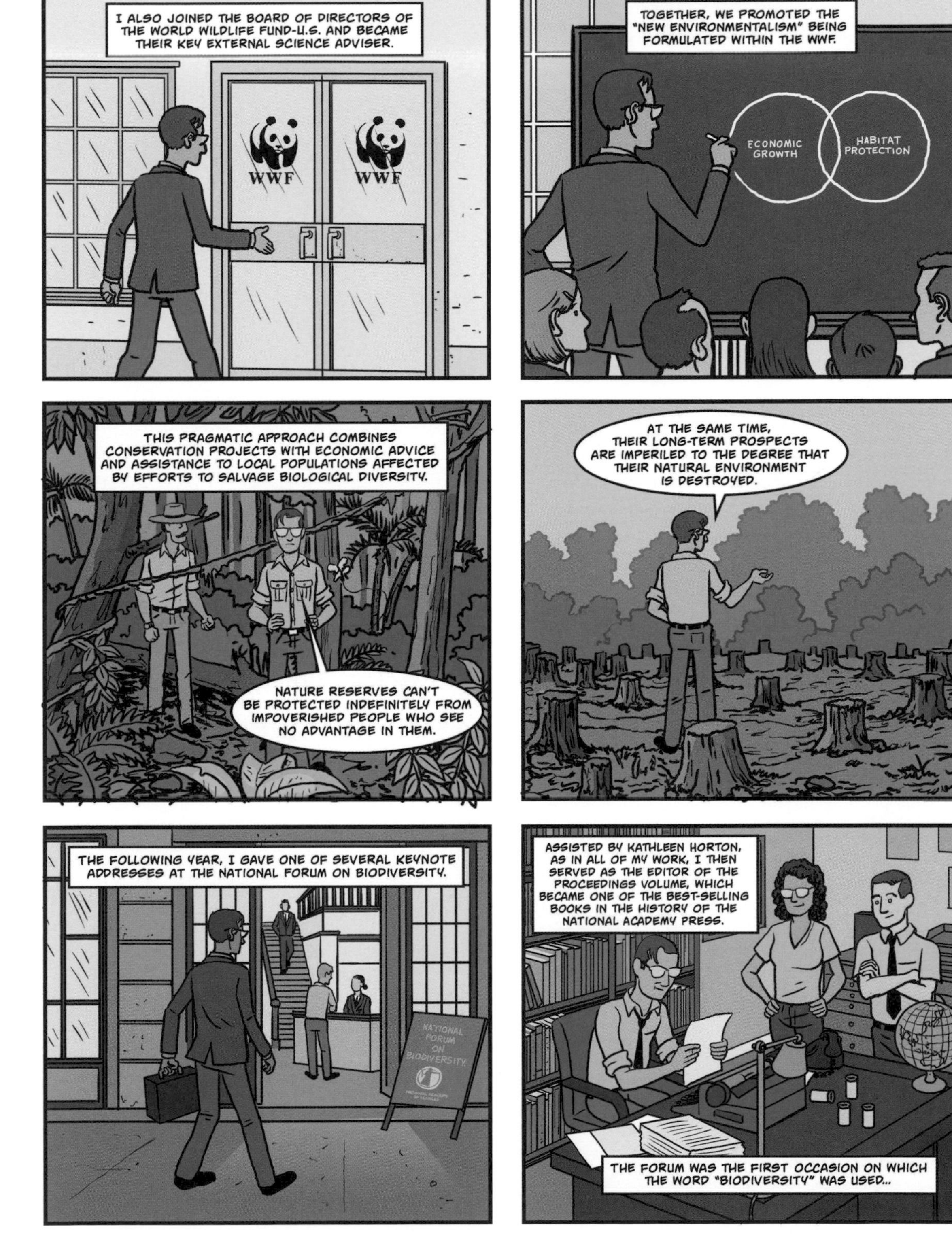
I ALSO JOINED THE BOARD OF DIRECTORS OF THE WORLD WILDLIFE FUND-U.S. AND BECAME THEIR KEY EXTERNAL SCIENCE ADVISER.
WWF
WWF
TOGETHER, WE PROMOTED THE "NEW ENVIRONMENTALISM" BEING FORMULATED WITHIN THE WWF.
ECONOMIC GROWTH
HABITAT PROTECTION
THIS PRAGMATIC APPROACH COMBINES CONSERVATION PROJECTS WITH ECONOMIC ADVICE AND ASSISTANCE TO LOCAL POPULATIONS AFFECTED BY EFFORTS TO SALVAGE BIOLOGICAL DIVERSITY.
NATURE RESERVES CAN'T BE PROTECTED INDEFINITELY FROM IMPOVERISHED PEOPLE WHO SEE NO ADVANTAGE IN THEM.
AT THE SAME TIME, THEIR LONG-TERM PROSPECTS ARE IMPERILED TO THE DEGREE THAT THEIR NATURAL ENVIRONMENT IS DESTROYED.
THE FOLLOWING YEAR, I GAVE ONE OF SEVERAL KEYNOTE ADDRESSES AT THE NATIONAL FORUM ON BIODIVERSITY.
NATIONAL FORUM ON BIODIVERSITY
ASSISTED BY KATHLEEN HORTON, AS IN ALL OF MY WORK, I THEN SERVED AS THE EDITOR OF THE PROCEEDINGS VOLUME, WHICH BECAME ONE OF THE BEST-SELLING BOOKS IN THE HISTORY OF THE NATIONAL ACADEMY PRESS.
THE FORUM WAS THE FIRST OCCASION ON WHICH THE WORD "BIODIVERSITY" WAS USED...

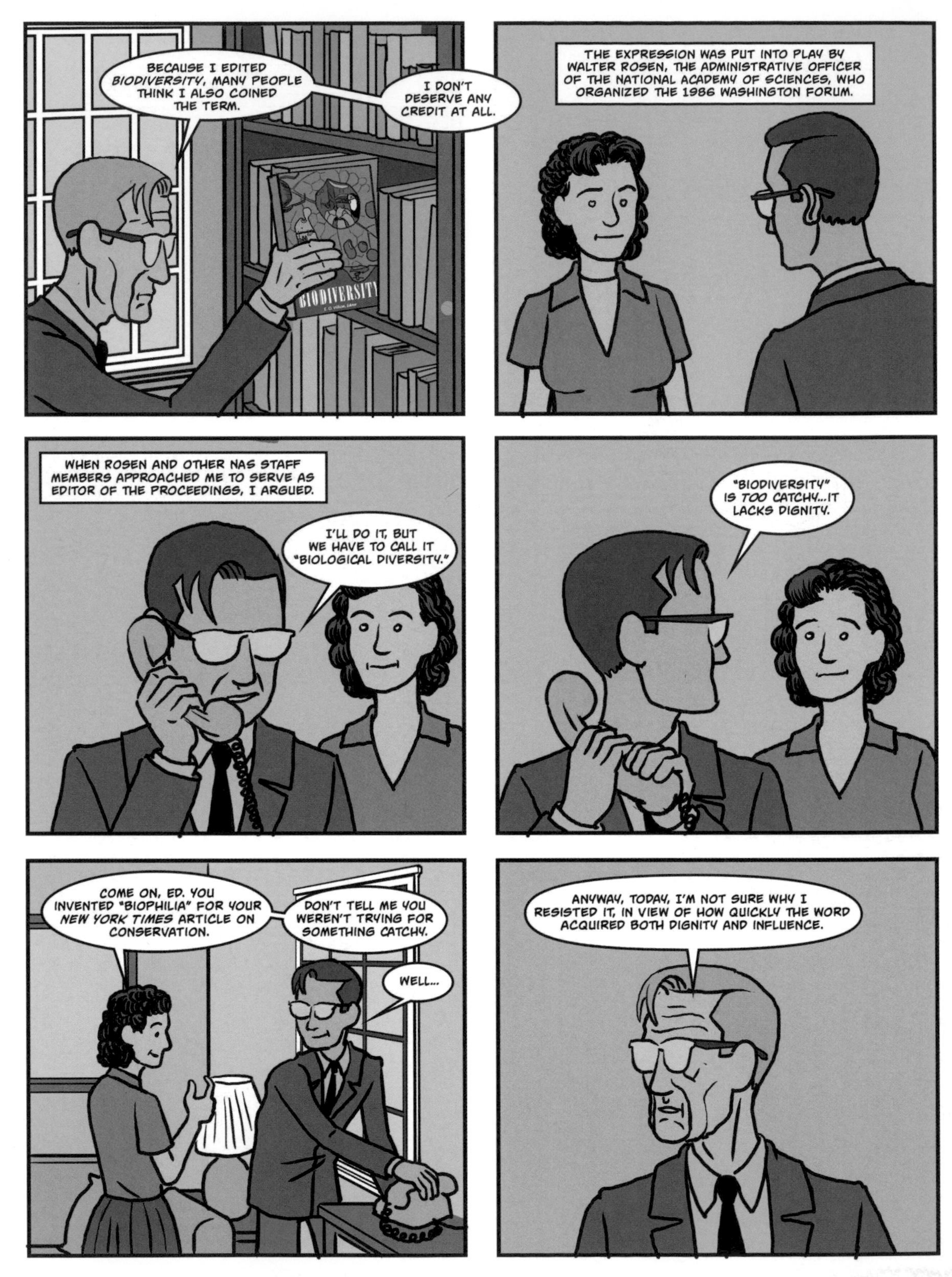
BECAUSE I EDITED *BIODIVERSITY*, MANY PEOPLE THINK I ALSO COINED THE TERM.
I DON'T DESERVE ANY CREDIT AT ALL.
BIODIVERSITY
THE EXPRESSION WAS PUT INTO PLAY BY WALTER ROSEN, THE ADMINISTRATIVE OFFICER OF THE NATIONAL ACADEMY OF SCIENCES, WHO ORGANIZED THE 1986 WASHINGTON FORUM.
WHEN ROSEN AND OTHER NAS STAFF MEMBERS APPROACHED ME TO SERVE AS EDITOR OF THE PROCEEDINGS, I ARGUED.
I'LL DO IT, BUT WE HAVE TO CALL IT "BIOLOGICAL DIVERSITY."
"BIODIVERSITY" IS *TOO* CATCHY...IT LACKS DIGNITY.
COME ON, ED. YOU INVENTED "BIOPHILIA" FOR YOUR *NEW YORK TIMES* ARTICLE ON CONSERVATION.
DON'T TELL ME YOU WEREN'T TRYING FOR SOMETHING CATCHY.
WELL...
ANYWAY, TODAY, I'M NOT SURE WHY I RESISTED IT, IN VIEW OF HOW QUICKLY THE WORD ACQUIRED BOTH DIGNITY AND INFLUENCE.

AS FOR BIOPHILIA, IT'S A DIFFERENT THING. ITS BASIC MANIFESTATION IS A PREFERENCE FOR CERTAIN NATURAL ENVIRONMENTS AS PLACES TO LIVE.
IN A PIONEERING STUDY OF THE SUBJECT, GORDON ORIANS, A ZOOLOGIST AT THE UNIVERSITY OF WASHINGTON, DIAGNOSED THE "IDEAL" HABITAT MOST PEOPLE PICK IF GIVEN A FREE CHOICE.
NEAR A LAKE, OCEAN, ETC.
ON A HILL, BLUFF, PROMINENCE
IN A PARK-LIKE ENVIRONMENT
CLOSE TO GROUND, WITH SMALL OR FINELY DIVIDED LEAVES
IT HAPPENS THAT THIS ARCHETYPE FITS A TROPICAL SAVANNA OF THE KIND PREVAILING IN AFRICA, WHERE HUMANITY EVOLVED FOR SEVERAL MILLIONS OF YEARS.

IS IT JUST A COINCIDENCE, THIS SIMILARITY?

CONSIDER A NEW YORK MULTIMILLIONAIRE WHO, WEALTHY ENOUGH TO LIVE ANYWHERE SHE WANTS, SELECTS A PENTHOUSE OVERLOOKING CENTRAL PARK.

IN A DEEPER SENSE THAN SHE PERHAPS UNDERSTANDS, SHE'S RETURNING TO HER ROOTS.

I ADMIT THAT BY THE ORDINARY STANDARDS OF NATURAL SCIENCE, THE EVIDENCE FOR BIOPHILIA REMAINS THIN.
AND MOST OF THE UNDERLYING THEORY OF ITS GENETIC ORIGIN IS HIGHLY SPECULATIVE.

STILL, THE LOGIC LEADING TO THE IDEA IS SOUND, AND THE SUBJECT'S TOO IMPORTANT TO NEGLECT.
IN MY OPINION, THE MOST IMPORTANT IMPLICATION OF AN INNATE BIOPHILIA IS THE FOUNDATION IT LAYS FOR AN ENDURING CONSERVATION ETHIC.

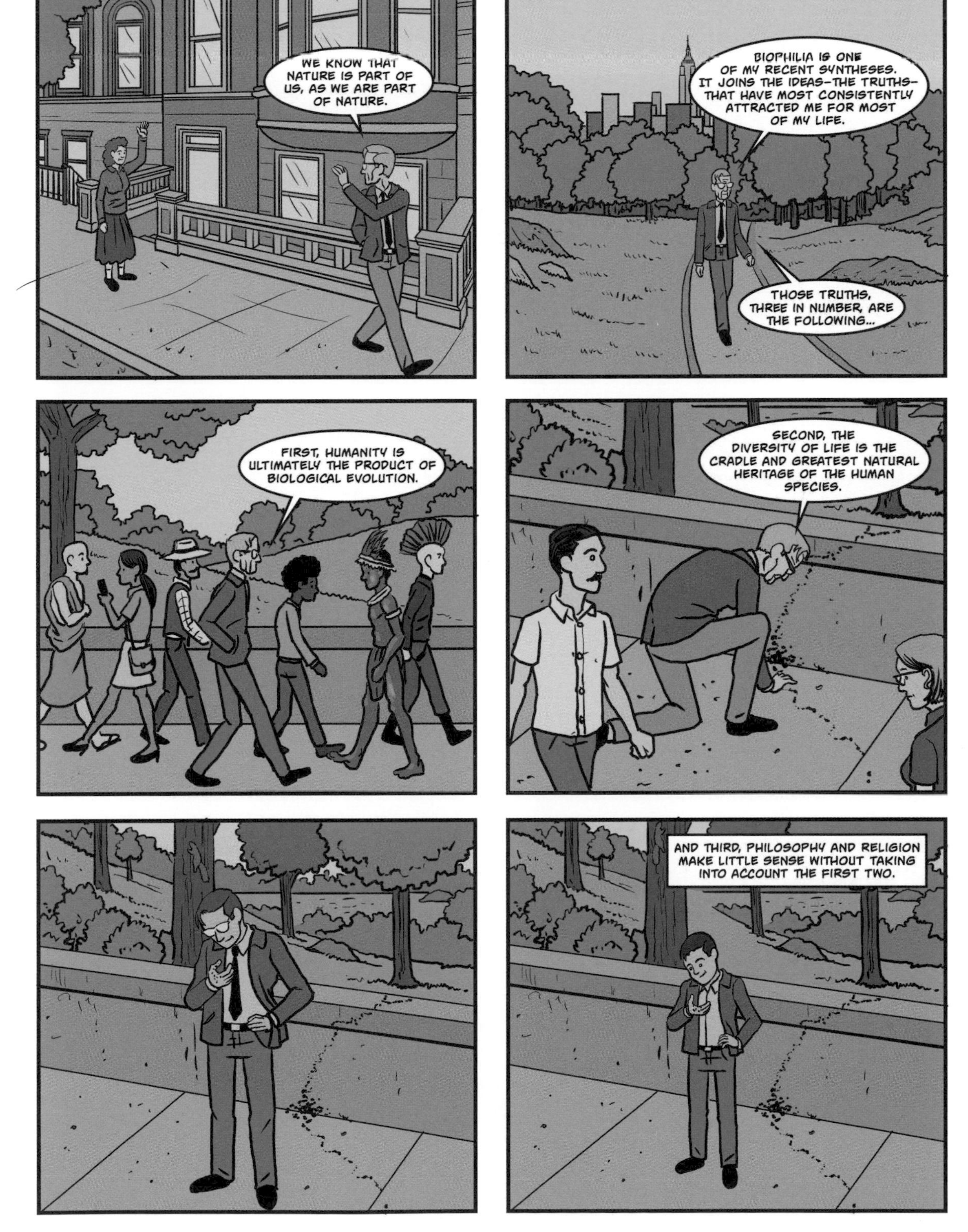
WE KNOW THAT NATURE IS PART OF US, AS WE ARE PART OF NATURE.
BIOPHILIA IS ONE OF MY RECENT SYNTHESES. IT JOINS THE IDEAS—THE TRUTHS—THAT HAVE MOST CONSISTENTLY ATTRACTED ME FOR MOST OF MY LIFE.
THOSE TRUTHS, THREE IN NUMBER, ARE THE FOLLOWING...
FIRST, HUMANITY IS ULTIMATELY THE PRODUCT OF BIOLOGICAL EVOLUTION.
SECOND, THE DIVERSITY OF LIFE IS THE CRADLE AND GREATEST NATURAL HERITAGE OF THE HUMAN SPECIES...
AND THIRD, PHILOSOPHY AND RELIGION MAKE LITTLE SENSE WITHOUT TAKING INTO ACCOUNT THE FIRST TWO.

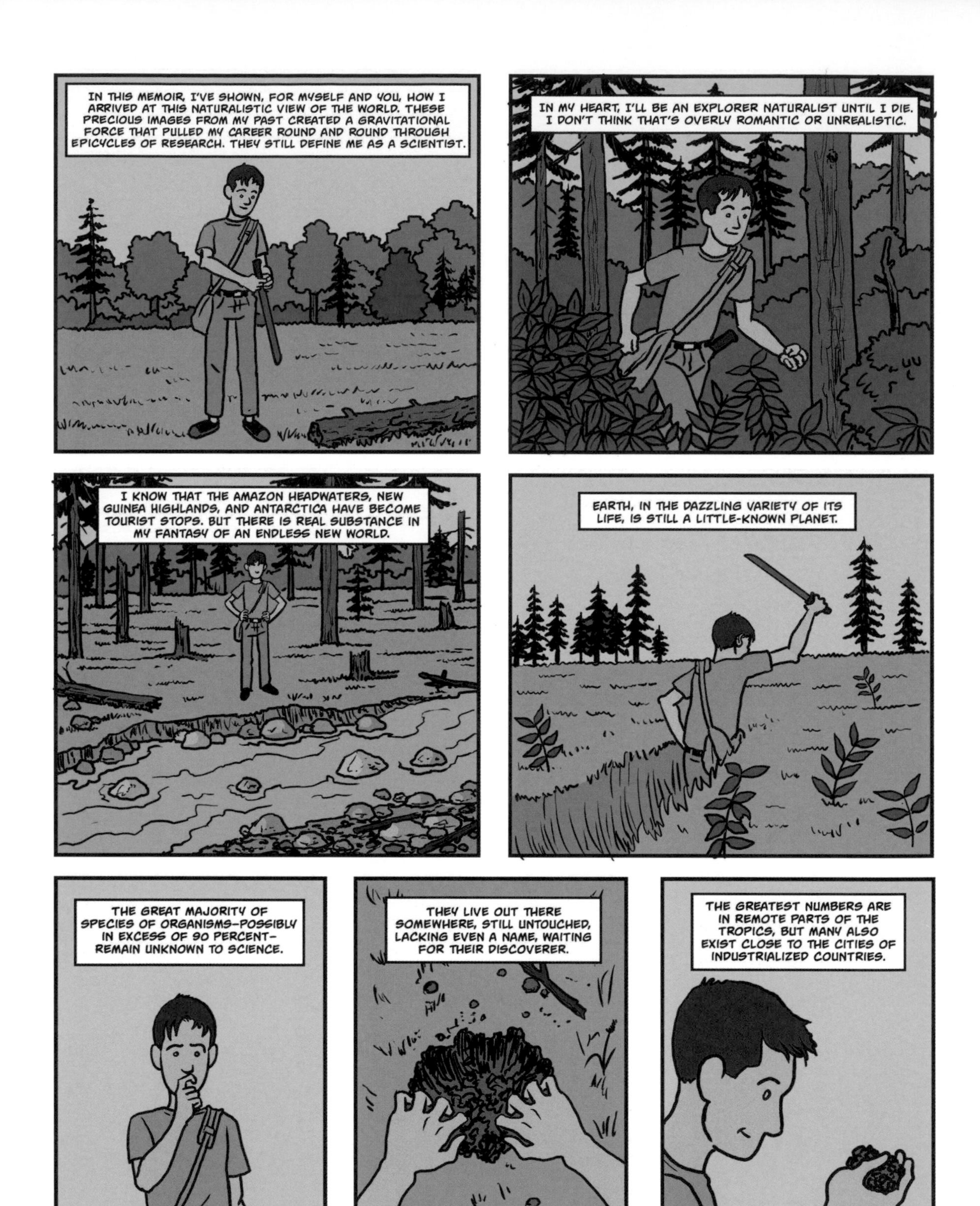
IN THIS MEMOIR, I'VE SHOWN, FOR MYSELF AND YOU, HOW I ARRIVED AT THIS NATURALISTIC VIEW OF THE WORLD. THESE PRECIOUS IMAGES FROM MY PAST CREATED A GRAVITATIONAL FORCE THAT PULLED MY CAREER ROUND AND ROUND THROUGH EPICYCLES OF RESEARCH. THEY STILL DEFINE ME AS A SCIENTIST.
IN MY HEART, I'LL BE AN EXPLORER NATURALIST UNTIL I DIE. I DON'T THINK THAT'S OVERLY ROMANTIC OR UNREALISTIC.
I KNOW THAT THE AMAZON HEADWATERS, NEW GUINEA HIGHLANDS, AND ANTARCTICA HAVE BECOME TOURIST STOPS. BUT THERE IS REAL SUBSTANCE IN MY FANTASY OF AN ENDLESS NEW WORLD.
EARTH, IN THE DAZZLING VARIETY OF ITS LIFE, IS STILL A LITTLE-KNOWN PLANET.
THE GREAT MAJORITY OF SPECIES OF ORGANISMS—POSSIBLY IN EXCESS OF 90 PERCENT—REMAIN UNKNOWN TO SCIENCE.
THEY LIVE OUT THERE SOMEWHERE, STILL UNTOUCHED, LACKING EVEN A NAME, WAITING FOR THEIR DISCOVERER.
THE GREATEST NUMBERS ARE IN REMOTE PARTS OF THE TROPICS, BUT MANY ALSO EXIST CLOSE TO THE CITIES OF INDUSTRIALIZED COUNTRIES.

IF I COULD DO IT ALL OVER AGAIN, AND RELIVE MY VISION IN THE TWENTY-FIRST CENTURY, I WOULD BE A MICROBIAL ECOLOGIST.
THE KEY TO TAKING THE MEASURE OF BIODIVERSITY LIES IN A DOWNWARD ADJUSTMENT OF SCALE. THE SMALLER THE ORGANISM, THE BROADER THE FRONTIER AND THE DEEPER THE UNMAPPED TERRAIN.
CONVENTIONAL WILDERNESSES OF THE OVERLAND TREK MAY INDEED BE GONE. MOST OF EARTH'S LARGEST SPECIES—MAMMALS, BIRDS, AND TREES—HAVE BEEN SEEN AND DOCUMENTED.
BUT MICROWILDERNESSES EXIST IN A HANDFUL OF SOIL OR AQUEOUS SILT COLLECTED ALMOST ANYWHERE IN THE WORLD. THEY, AT LEAST, ARE CLOSE TO A PRISTINE STATE AND STILL UNVISITED.
MINUTE CREATURES SWARM AROUND US, AN ANIMATE MATRIX THAT BINDS EARTH'S SURFACE.
TEN BILLION OF THEM LIVE IN A GRAM OF ORDINARY SOIL.
THEY REPRESENT THOUSANDS OF SPECIES, ALMOST NONE OF WHICH ARE KNOWN TO SCIENCE.
THEY ARE OBJECTS OF POTENTIALLY ENDLESS STUDY AND ADMIRATION.
INTO THAT WORLD I WOULD GO WITH THE AID OF MODERN MICROSCOPY AND MOLECULAR ANALYSIS.
I WOULD CUT MY WAY THROUGH CLONAL FORESTS SPRAWLED ACROSS GRAINS OF SAND...

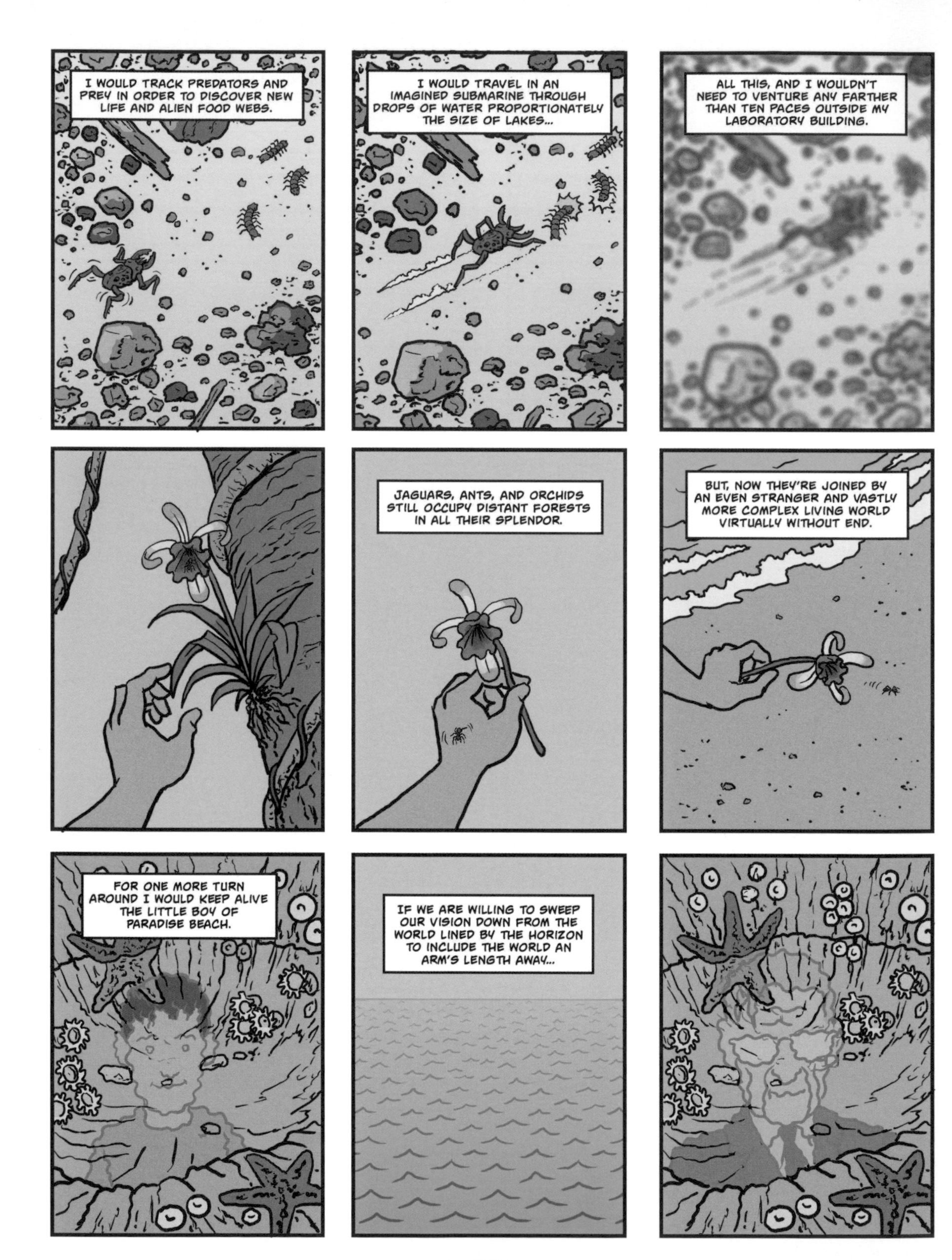
I WOULD TRACK PREDATORS AND PREY IN ORDER TO DISCOVER NEW LIFE AND ALIEN FOOD WEBS.
I WOULD TRAVEL IN AN IMAGINED SUBMARINE THROUGH DROPS OF WATER PROPORTIONATELY THE SIZE OF LAKES...
ALL THIS, AND I WOULDN'T NEED TO VENTURE ANY FARTHER THAN TEN PACES OUTSIDE MY LABORATORY BUILDING.
JAGUARS, ANTS, AND ORCHIDS STILL OCCUPY DISTANT FORESTS IN ALL THEIR SPLENDOR.
BUT, NOW THEY'RE JOINED BY AN EVEN STRANGER AND VASTLY MORE COMPLEX LIVING WORLD VIRTUALLY WITHOUT END.
FOR ONE MORE TURN AROUND I WOULD KEEP ALIVE THE LITTLE BOY OF PARADISE BEACH.
IF WE ARE WILLING TO SWEEP OUR VISION DOWN FROM THE WORLD LINED BY THE HORIZON TO INCLUDE THE WORLD AN ARM'S LENGTH AWAY...

...WE CAN SPEND A LIFETIME IN
A MAGELLANIC VOYAGE
AROUND THE TRUNK
OF A SINGLE
TREE.

About Edward O. Wilson

REGARDED as one of the world's preeminent biologists, Edward O. Wilson grew up in south Alabama and the Florida Panhandle, where he spent his boyhood collecting snakes, butterflies, and ants—and more than a few comic books.

Throughout his long career, Wilson made seminal contributions to the study of evolution and ecology, created the field of sociobiology, and was a pioneer in efforts to preserve and protect Earth's biodiversity. He is the author of more than thirty books, including *The Ants* and *On Human Nature*, both of which were awarded the Pulitzer Prize. Wilson is currently Faculty Emeritus in the Museum of Comparative Zoology and Pellegrino University Professor, Emeritus at Harvard University.

About the Authors

C.M. Butzer is an illustrator, printmaker, and cartoonist. His work has appeared in numerous books, publications, and textiles. In 2009, he wrote and drew *Gettysburg: The Graphic Novel* for HarperCollins. Butzer is also a storyboard and concept artist whose clients have included Facebook, Google, and Microsoft. After 13 years of residing in Brooklyn, he and his wife moved to Hilo, Hawaii, where he does not miss winter.

Jim Ottaviani has written a dozen (and counting) graphic novels about scientists. His most recent books are *Astronauts: Women on the Final Frontier*, *Hawking*, *The Imitation Game* (a biography of Alan Turing), *Primates* (about Jane Goodall, Dian Fossey, and Biruté Galdikas) and *Feynman*. His books are *New York Times* bestsellers, have been translated into over a dozen languages, and have received praise from publications ranging from *Nature* and *Physics World* to *Entertainment Weekly* and *Variety*. Jim lives in Michigan and comes to comics via careers in nuclear engineering and librarianship.

Hilary Sycamore didn't know there was such a thing as a colorist until 20 years ago when she began coloring a graphic novel to help a friend. She was hooked right away—and hasn't stopped since! Originally from London, England, she holds a B.A. in fine art from the University of London and taught art and design to all grades until her family moved to the United States in 2000. She has colored a wide range of projects, including *Spill Zone* by Scott Westerfeld and Alex Puvilland, *Mighty Jack* by Ben Hatke, and *Feynman* by Jim Ottaviani and Leland Myrick.

ISLAND PRESS, a nonprofit publisher, provides the best ideas and information to those seeking to understand and protect the environment and create solutions to its complex problems.

Working with leading thinkers from around the world, Island Press elevates voices of change, shines a spotlight on crucial issues, and focuses attention on sustainable solutions.

Island Press gratefully acknowledges major support from The Bobolink Foundation, Caldera Foundation, The Curtis and Edith Munson Foundation, The Forrest C. and Frances H. Lattner Foundation, The JPB Foundation, The Kresge Foundation, The Summit Charitable Foundation, Inc., and many other generous organizations and individuals.

Generous support for the publication of this book was provided by the Alfred P. Sloan Foundation, Margot and John Ernst, Pamela and Byrne Murphy, Georgia Chafee Nassikas, and many other generous individuals.

The opinions expressed in this book are those of the author(s) and do not necessarily reflect the views of our supporters.